Learning about materials

ROYAL SOCIETY OF CHEMISTRY

Learning about materials

Edited by Ted Lister and Colin Osborne

Designed by Imogen Bertin, Sara Roberts and Grace Donnelly

Published by the Education Division, The Royal Society of Chemistry

Printed by The Royal Society of Chemistry

For further information on other educational activities undertaken by The Royal Society of Chemistry write to:

The Education Department
The Royal Society of Chemistry
Burlington House
Piccadilly
London W1V OBN

ISBN 0–85404–920–7

British Library Cataloguing in Data.
A catalogue for this book is available from the British Library.

RS•C

Introduction

This publication is the outcome of a new venture for The Royal Society of Chemistry in conjunction with The Institute of Materials; and the Armourers' and Brasiers' company Learning Material Workshops . Three workshops were held in 1995:

▼ at Johnson Matthey, Reading on the extraction of platinum group metals;

▼ at Raychem, Swindon on "smart materials"; and

▼ at Chemoxy International, Middlesborough on "biodiesel".

A different group of chemistry teachers was involved in each. Each group spent a day with one of the companies and was given a presentation by the company on some aspect of their work. This was followed by a brainstorming session to suggest ideas for useful learning material based on what they had seen and heard. The following day was spent drafting material which was then edited and, in most cases trialled. This book is the result.

The pack contains teacher's notes and material to photocopy for use by students. An extra copy of this material is supplied as separate masters for easy photocopying.

Acknowledgements

The Royal Society of Chemistry thanks the three companies involved and, in particular, David Boyd of Johnson Matthey, Narinder Kehal of Raychem and David Randall of Chemoxy International who gave freely of their time and expertise both during the workshops and afterwards. The Society also thanks Johnson Matthey for financial assistance in producing the video.

This page has been intentionally left blank.

RS•C

Platinum extraction

Contents

RS•C

Introduction

This pack is based on platinum and its extraction. The aim of the units is to provide some interesting and relevant material for pre-16 science/chemistry students and post-16 chemistry students.

A video (*Platinum, the noble metal*; VHS approx 15 minutes) can be used with this pack and supports some of the activities. It covers the history, geology, mining, extraction and uses of platinum. All of the activities can be done without the video.

The pack is in three parts. Each part is independent of the others.

The first part *The platinum story* gives the broad picture of how platinum – and other associated rare metals – are mined and refined. It could be used at all levels as an introduction, though it is aimed at pre-16 courses.

The second part *The isolation of the platinum group metals (PGM)*, which includes a short practical demonstration, and the third *Aspects of the platinum group metals (PGM)* are suitable for post-16 courses.

This pack is the result of a Royal Society of Chemistry visit to Johnson Matthey Technology Centre near Reading, December 1995, and was developed in collaboration with Johnson Matthey plc and Anglo American Platinum Corporation. It is based upon original ideas produced by :

▼   Isobel Crocker, The Dame Alice Harper School;

▼   David Francis , Portsmouth College;

▼   Andrew Hayes, City of Bath College of Further Education;

▼   Lesley Stanbury, St. Albans School;

▼   Ian Thomas, Ashlyn's School; and

▼   Diana Townsend, Ballard College.

RS•C

RS•C

Part 1 Pre-16 – The platinum story

Teacher's notes

This section includes information on platinum (which can be used along with the video), questions to test comprehension, a word search and a hidden word puzzle.

Curriculum links

Useful products from metal ores and rocks, periodicity, properties of metals, transition metals.

Level

Pre-16 science/chemistry courses

Timing

60–70 mins

This pack contains:

▼   an information sheet, *The platinum story*, on the extraction of platinum;

▼   a question sheet, *Platinum question sheet*, relating to the above information and/or to the video;

▼   an alternative simpler word completion exercise, *Platinum comprehension*, with straightforward questions relating to some text on the extraction of platinum;

▼   a puzzle sheet – *Precious metal puzzle*; and

▼   a *Wordsearch* with questions.

All of the above are available as photocopy masters.

Possible lesson plan

To introduce the topic, ask the class to name a metal comparable in price to gold – hopefully someone will suggest platinum and be aware of its use in jewellery. Begin a discussion on the possible reasons for platinum being so expensive. This might encompass useful properties (unreactive and high melting point), difficulty in extraction, scarcity *etc*. This last point could be emphasised by pointing out that a handful of typical platinum ore contains so little platinum that it is worth less than 10p.

Now show the video (and/or give out the information sheet) followed by the question sheet, *Platinum question sheet*. If you use the information sheet, which is much more detailed, it may be useful to demonstrate a precipitation reaction in a test–tube. For example, adding copper(II) sulfate solution to sodium carbonate solution makes clear the idea of separating out by precipitation.

Several of the questions in part B of the question sheet are more open-ended and could be used to stretch the more able students. One or two of these questions could be chosen for students to write up answers in the normal way, or to give a presentation to the class.

The simpler sheet may be used, if more appropriate, to accompany the video. The *Precious metal puzzle* sheet and *Wordsearch* are available for further work.

RS•C

Answers to the platinum question sheet

Part A

1. a) South Africa.

   b) It is rare and has many uses.

   c) Three tonnes.

2. a) The main steps:

   Mining – the ore is brought to surface.

   Crushing – the ore is first crushed and then a wet slurry is formed in the ball mills.

   Froth flotation – a detergent is added and the metal ore particles float on the foam where they are separated off.

   Smelting – the ores are melted, driving off some sulfur, to leave metals and metal sulfides.

   Separation

   (i) By using a magnetic drum which separates the metals (because nickel is magnetic and is present in the alloy of platinum group metals (PGM)) from the sulfides

   (ii) Final separation of PGM using solvents or precipitation.

   b) Sulfur dioxide is converted to sulfuric acid.

3. Cobalt (Co); copper (Cu); gold (Au); iridium (Ir); nickel (Ni); osmium (Os); palladium (Pd); rhodium (Rh); ruthenium (Ru); silver (Ag).

Part B

1.

Jewellery	– stays shiny because it is unreactive
Catalytic converter	– suitable catalytic properties
Electronics	– good conductor and does not corrode
Fuel cells	– chemically inert (catalytically active) electrodes
Chemotherapy	– some platinum compounds are anticancer agents
Glass fibre drawing	– high melting point and does not corrode

RS•C

RS•C

2.   Factors which might affect the price include:

More expensive	Cheaper
More uses found for platinum Resources used up Mining becomes more difficult as resources dwindle	Alternative cheaper material found, suitable for same use New deposits found Better mining design Recycling made easy

3.   Any suitable research answer.

4.   Catalytic converters change carbon monoxide, unburnt hydrocarbons and nitrogen oxides into carbon dioxide, water and nitrogen.

      Lead compounds poison the catalyst.

      Despite difficulties in collecting the converters from scrapped cars, an increasing proportion is being recycled, particularly in North America.

Platinum comprehension answers

1.   Metals, Periodic, ore, solvent, gold, catalyst.

2.   South Africa.

3.   Two thousand million (two billion) years ago.

4.   It is rare and has many uses.

5.   It converts harmful exhaust gases to safer gases.

Precious metal puzzle answers

1.   Nickel.

2.   Platinum.

3.   Smelting.

4.   Exhaust.

5.   Palladium.

6.   Merensky.

7.   Corrosion.

8.   South Africa.

9.   Steel.

Hidden word: catalysts

Wordsearch answers

a)   Iron, cobalt.

b)   Silver.

RS•C

Part 2 Post-16 – The isolation of the platinum group metals (PGM)

Teacher's notes

This is a comprehension exercise on the solvent extraction of platinum, supported by a demonstration of the partition of iodine between two solvents. Students will probably require a brief description of the overall process of extracting platinum before they read the passage *Solvent extraction of platinum*. Alternatively, they could read *The platinum story,* or watch the video if available.

Curriculum links

Equilibrium; partition coefficients; revision of amines.

Timing

60 –70 mins

Level

Post-16 chemistry courses

Apparatus

- ▼ Two boiling tubes with bungs
- ▼ Rack for boiling tubes
- ▼ Dropping pipette.

Chemicals

- ▼ 100 cm^3 hexane
- ▼ 100 cm^3 1 mol dm^{-3} aqueous potassium iodide solution
- ▼ A few small crystals of iodine.

Safety

- ▼ Wear eye protection
- ▼ Hexane is flammable
- ▼ Iodine is harmful by skin contact and gives off a toxic vapour that is dangerous to the eyes
- ▼ It is the responsibility of the teacher to carry out a risk assessment.

RS•C

Possible lesson plan

1. Introduce the topic by showing the video (if available) and summarise with the following OHP showing the flow chart of platinum extraction.

Flow chart of platinum extraction as a whole

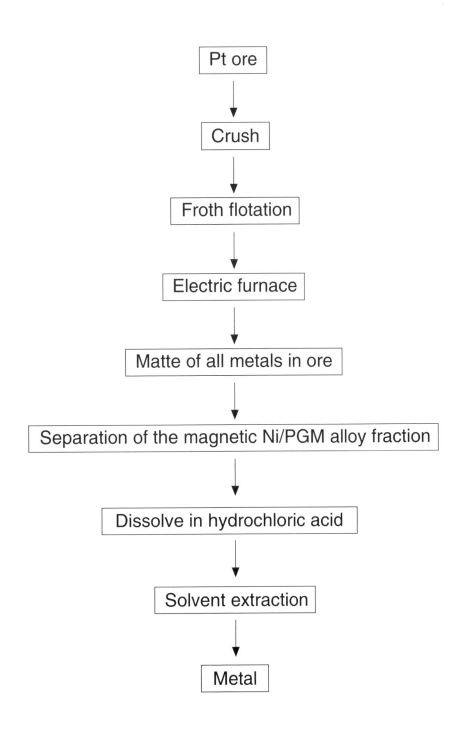

RS•C

2.   Demonstrate the nature of differential solubility in two immiscible solvents by showing how iodine distributes between an aqueous solvent and an organic solvent. This could alternatively be done as a class experiment.

Practical details

Use boiling tubes and bungs

a)   To demonstrate the colour in each solvent:

Boiling tube 1
Dissolve one small crystal of iodine in a 2 cm depth of 1 mol dm^{-3} aqueous potassium iodide solution. A yellow/brown solution forms.

Boiling tube 2
Dissolve one small crystal of iodine in a 2 cm depth of hexane. A pink/purple solution forms.

b)   To demonstrate partition:

To boiling tube 1, add 2 cm depth of hexane and shake well.

What do you see? Discuss the partition of iodine between the two solvents.

Optional

Remove the organic layer using a dropping pipette and add a further 2 cm depth of hexane and shake.

Mention that the organic solvents used in the extraction of platinum contain dissolved amines. (The platinum metal is recovered finally from the organic solvent by precipitation via an aqueous solvent, and then strongly heating the solid compound to leave platinum metal.)

3.   Use the worksheet *Solvent extraction of platinum*.

Background notes for teachers on solvent extraction

The platinum group of metals consists of platinum, palladium, rhodium, iridium, ruthenium and osmium. They may be separated from each other (after extraction from their ores) by precipitation reactions. An alternative to this is **solvent extraction.**

The platinum group metals (PGM) occur in minute quantities in deposits of copper-nickel sulfide ore. This material is mined commercially in a number of places in the world, but the three principal areas are in Canada, countries of the former USSR and South Africa (in the Merensky Reef, Transvaal). The Merensky deposits, though the major source, yield less than 10 grams of platinum from three tonnes of ore.

The crushed ore is concentrated by:

▼   **physical beneficiation**, (a technical term for the physical processes of crushing, sieving and flotation);

▼   **pyrometallurgical** techniques (the smelting processes); and

▼   **hydrometallurgical** techniques (the dissolving processes).

This eventually yields a concentrate containing around 50% PGM by mass. The remainder is largely gold, silver, copper, nickel and other base metals. The PGM are traditionally separated from one another and the other metals by a series of selective

RS•C

precipitation techniques. These are generally inefficient in terms of the degree of separation achieved, and usually the precipitate has to be dissolved again and reprecipitated at least once more to get a pure product. Even when it proves possible to remove the desired elements completely from solution, the precipitate requires thorough washings to remove contamination from the original solution.

In the past few years the application of solvent extraction techniques has been used as a more efficient alternative for the separation of the PGM.

Once the initial capital investment in equipment has been made, solvent extraction processing has considerable advantages over the conventional precipitation processes. These include:

▼ greater safety because the platinum solutions (which have some allergenic properties) are in a closed system;

▼ reduced overall processing time – particularly for the major elements (*eg* platinum);

▼ improved yield from fewer steps; and

▼ the solvent can be reused.

In solvent extraction the base metals, such as iron, must be removed first because they tend to behave in a similar way to the PGM and are therefore more difficult to separate later on.

For a particular batch of ore, the steps involved in solvent extraction run continuously (see overleaf) and can be controlled automatically using sophisticated analytical instrumentation. Consequently, labour costs are reduced compared with a conventional batch process.

RS•C

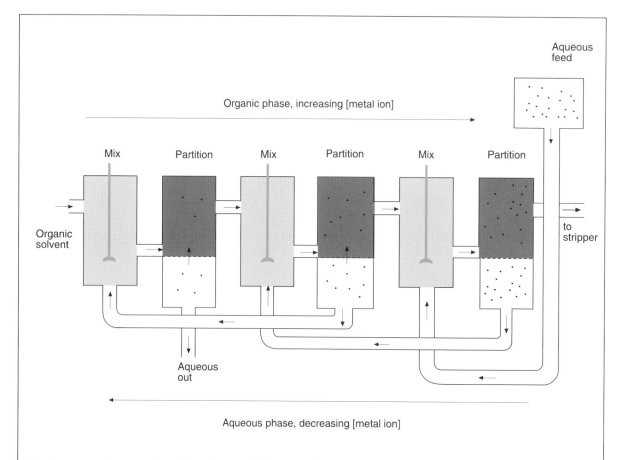

A schematic diagram showing the principle of continuous counter current solvent extraction. Organic solvent moves from left to right through a series of mixers and separating tanks, picking up metal ions as it does so. The aqueous phase moves from right to left, losing metal ions as they are extracted into the organic solvent. The whole process has the same effect as a large number of successive batch solvent extractions. The metal ions are then stripped from the organic solvent which is recycled.

Solvent extraction

RS•C

Answers to questions on the solvent extraction of platinum

1. (a) (i) Upper layer.

   (b) (i) Organic upper layer.

   (ii) Iodine is a molecular compound therefore, as "like dissolves like", a covalently bonded molecule is likely to dissolve to a greater extent in the organic layer.

   (iii) Greater than one.

2. a) Tertiary b) secondary c) quaternary d) primary

3. $RNH_2 + HCl \rightarrow RNH_3^+Cl^-$

4. a) Quaternary ammonium.

   [The relative efficiency of extraction of $[PtCl_6]^{2-}$ by the different classes of amines is :

   quaternary > tertiary > secondary > primary]

   b) Partition coefficient 1.0 implies equal distribution of solute between the organic and aqueous layers.

   c) Primary and secondary amines.

RS•C

Part 3 Post-16 –
Aspects of the platinum group metals (PGM)

Teacher's notes

This worksheet can stand alone as a teaching and comprehension exercise. It is based on the principles by which the PGM are extracted and separated from a solution of the PGM in 6 mol dm^{-3} hydrochloric acid.

The reading exercise is particularly useful to stretch more able students although the questions are relatively straightforward.

Curriculum links

d-Block elements and ligands, complexes, oxidation numbers, shapes of complexes, redox reactions and use of $E^\ominus$.

Level

Post-16 courses.

Timing

60–70 mins.

Answers to questions on aspects of the platinum group metals (PGM)

1. a)   Molecules or ions which form dative bonds to transition metal ions.

   b)   d-Block metal surrounded by ligands.

   c)   The number of ligands which surround the metal.

   d)   A complex in which two metal atoms are joined via an oxygen atom.

$$eg \quad - \overset{\displaystyle |\;\diagup}{\underset{\diagup\;|}{M}} - O - \overset{\displaystyle |\;\diagup}{\underset{\diagup\;|}{M}} -$$

2.   One example from a), b) and c) below

   a) $PdCl_4^{2-}$

         $PtCl_4^{2-}$

   b) $[RuCl_6]^{3-}$

         $[RuCl_5(H_2O)]^{2-}$

         $[RuCl_4(H_2O)_2]^-$

         $[RuCl_3(H_2O)_3]$

         $[OsCl_6]^{3-}$

         $[OsCl_5(H_2O)]^{2-}$

         $[OsCl_4(H_2O)_2]^-$

$[RhCl_6]^{3-}$

$[RhCl_5(H_2O)]^{2-}$

$[RhCl_4(H_2O)_2]^-$

$[IrCl_6]^{3-}$

$[IrCl_5(H_2O)]^{2-}$

$[IrCl_4(H_2O)_2]^-$

$[AuCl_4]^-$

c) $[RuCl_6]^{2-}$

$[Ru_2OCl_{10}]^{4-}$

$[Ru_2OCl_8(H_2O)_2]^{2-}$

$[OsCl_6]^{2-}$

$[IrCl_6]^{2-}$

$[PdCl_6]^{2-}$

$[PtCl_6]^{2-}$

3.   a)   (i) Octahedral.

(ii) Two octahedra joined by an oxygen bridge as in 1(d).

b)

(cis and trans)

(cis and trans)

(mer and fac)

4.   a) $[RhCl_5(H_2O)]^{2-} > cis\ [RhCl_4(H_2O)_2]^- > fac\ [RhCl_3(H_2O)_3] > [RhCl_6]^{3-}$

b) $[RhCl_5(H_2O)]^{2-} > [RhCl_6]^{3-} > cis\ [RhCl_4(H_2O)_2]^- > fac\ [RhCl_3(H_2O)_3]$

5.   a) $[MCl_4]^- > [MCl_6]^{2-}$ because for steric reasons it is harder to pack two bulky mono cations around $[MCl_6]^{2-}$ than one around $[MCl_4]^-$.

NB In the organic phase, the species exist as ion pairs with cations packing around the anions. In the aqueous phase, the cations and anions are separate.

b) $[MCl_6]^{2-} > [MCl_5(H_2O)]^{2-}$ due to hydrogen bonding of the water ligand with the water molecules, this complex tends to stay in the aqueous (acid) phase, making it less likely to dissolve in the organic phase.

Information sheets – the platinum story

The platinum story

Platinum has been valued as a jewellery metal since ancient times. It was used by the Egyptians over 2000 years ago. It was thought the metal was a form of silver and the name platina means little silver.

Platinum is a rare metal and its ores are found only in a few places around the world mainly in igneous rock deposits formed some two thousand million years ago. South Africa produces the largest amount of platinum in the world. The Merensky Reef deposits in the Transvaal (see Map) cover a wide area and are some of the richest in the world. Even so the deposits produce only 10 grams of platinum for every three tonnes of rock that are mined.

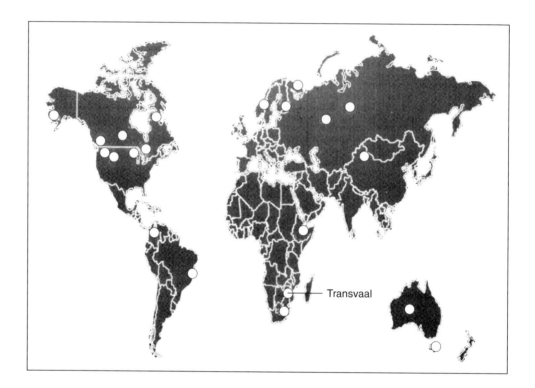

Transvaal

The location of platinum deposits

Platinum is found in ores containing other metals, often base metals - copper, nickel and cobalt – but also other very rare metals. **Iridium, osmium, palladium, platinum, rhodium and ruthenium** are called the platinum group metals (PGM). Gold and silver are also sometimes present in smaller quantities, so the ore is very precious.

Platinum and the other rare metals are very unreactive and have high melting points. Their many uses (apart from jewellery) include applications in electronics, chemotherapy, glass fibres (where platinum is used to make the spinneret through which molten glass is drawn) and in fuel cell technology as catalysts. They are useful as catalysts in many processes. Platinum is part of the catalyst system in the catalytic converters of car exhausts.

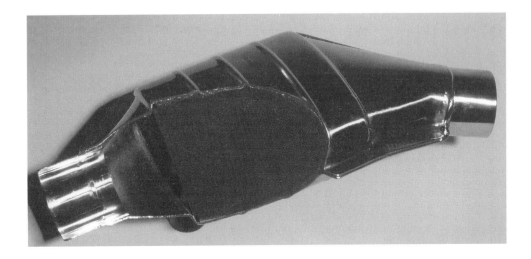

Inside the catalytic converter is a ceramic honeycomb coated with small quantities of platinum group metals

From ore to platinum

The seams of the ore-bearing rock are mined underground and large lumps of rock are brought to the surface.

Platinum ore is mined up to 1.5 km below the surface

These lumps of rocks are first crushed between steel jaws and then mixed with water and put into ball mills, which are rotating steel drums partly filled with steel balls. These grind the ore to a wet slurry.

Ball mills – giant rotating drums

The slurry is mixed with a special detergent which produces a froth. The metal compounds contained in the rocks are attracted to the surface of the foam and can be removed. This is called froth flotation.

Froth flotation

The next step is smelting. The metal compounds are put in an electric furnace at a temperature of 1400 °C and processed to drive off some of the sulfur from the sulfide ores as sulfur dioxide.

Smelting

Any sulfur dioxide produced from the sulfides is converted to sulfuric acid leaving a liquid called a **matte** which is cooled to a solid. The solid is a mixture of:

▼   nickel and copper sulfides; and

▼   a metallic alloy (a mixture of metals) consisting of the PGM enriched with metallic nickel.

This solid is crushed to a powder. Then it is separated into the parts described above by using a magnetic drum. This works because nickel is a magnetic metal so the alloy part of the powder, which contains the PGM and nickel, sticks to the drum and is separated from the non-magnetic sulfides.

Solidified matte

The non-magnetic sulfides are next dissolved in sulfuric acid. This leaves behind a valuable slurry consisting of any precious metals not picked out by the magnetic drum. The base metals such as copper and nickel are recovered from the solution by electrolysis.

The magnetic mixture, containing most of the PGM, is dissolved in hydrochloric acid ready for the final stage.

The PGM, now present as chlorides, are separated from each other using some rather complex processes. For example, a solution can be added which causes one of the metals to form an insoluble solid salt which can be filtered off (this is called precipitation), and separates this one from the others. Alternatively, some can be separated by a process called solvent extraction where the metal compounds dissolve to different extents in different solvents and are separated from each other one after another.

Platinum question sheet

Part A

Extraction

1. a) Where in the world are the largest deposits of platinum found?

    b) Why is platinum so expensive?

    c) How much ore do you need to extract 10 g of platinum?

2. a) List the main stages used in extracting platinum.
        Outline what happens at each stage (a flow diagram may help).

    b) The platinum ore contains large amounts of sulfur. This could result in large amounts of sulfur dioxide being released into the atmosphere. What is done to prevent this?

3. Other metals are produced at the same time as platinum. Give the name and the chemical symbol of each of these metals.

Part B

Uses

1. List as many items as you can that are made of platinum. For each item say why platinum is suitable.

2. Imagine that you are asked to forecast whether platinum will still be as valuable in 20 years time as it is today.
    What factors might make it more expensive?
    What factors might make it cheaper?
    Try to make a reasoned case to back up your forecast.

3. Write a report on the properties and uses of one of the other metals extracted at the same time as platinum.

4. Find out about catalytic converters in cars.
    What is their purpose?
    Why must they be used with lead-free petrol?
    What happens to the catalyst when a car is scrapped?

Platinum comprehension

We are all familiar with metals such as iron and aluminium which we see and use every day. Enormous amounts of these metals are extracted from their ores which generally contain quite a lot of the metal.

Platinum, on the other hand, is found in tiny amounts in deposits of ore mixed up with even smaller amounts of other similar metals, such as rhodium and palladium.

The ore is found only in a few places in the world, mainly in igneous rock deposits which are approximately two thousand million years old. The Merensky Reef deposits in the Transvaal, South Africa cover a wide area but produce only 10 g of platinum for every three tonnes of rock that is mined.

Separating the platinum out is a very complicated process with lots of stages, using unusual methods such as solvent extraction.

Platinum has always been valued as jewellery since ancient times and was known for example in Egypt over 2000 years ago.

Recently demand for platinum and other similar metals has increased due to their use as catalysts in the exhaust systems of cars.

Questions

1. Copy and complete the passage below using the list of words provided.

    catalyst   gold   metals   ore   Periodic   solvent

    Platinum belongs to a set of _ _ _ _ _ _ _ _ in the _ _ _ _ _ _ _ _ Table known as the platinum group metals (PGM). It is found in tiny amounts in deposits of _ _ _ _ _ _ _ _, in which it is mixed up with even smaller amounts of other similar metals such as rhodium. Getting the metal out is a very difficult process, using methods such as _ _ _ _ _ _ _ _ extraction. Platinum is a rare and expensive metal which is even more valuable than _ _ _ _ _ _ _ _. It is widely used in the exhaust systems of motor cars as a _ _ _ _ _ _ _ _.

2. Where is the Merensky Reef?

3. How long ago were the platinum deposits formed?

4. Why do you think platinum is so valuable?

5. What does a catalyst in a car exhaust do?

Precious metal puzzle

Complete the puzzle by filling in the blanks 1–9. When you have finished, you will find the word between the stars. It reveals what many transition metals are used for. You may find a Periodic Table useful.

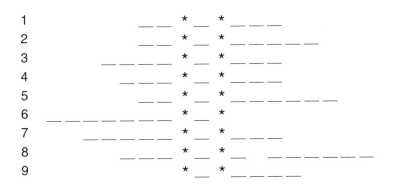

1. A metal with 28 protons per atom extracted at the same time as platinum.

2. An expensive metal used in jewellery.

3. Means heating an ore so it reacts and melts, giving free metal.

4. Pt and Pd are used to 'clean up' gases from a car .... pipe.

5. A useful metal with catalytic power which has the same name as a theatre in London. [Hint – its symbol is Pd].

6. Name of 'reef' where most of the world's platinum is found..

7. An oxidation problem affecting most metals but not platinum!

8. A country rich in several deposits of minerals which contain precious metals.

9. A commonly used alloy of element with atomic number 26.

Wordsearch

Platinum is the most important metal in the set of metals which are called the platinum group metals (PGM). (Note that the word group is not being used in the strict sense in which we use it in the Periodic Table.) They are usually found together in rocks where they are present as minerals. Find the six metals **PLATINUM, PALLADIUM, RHODIUM, IRIDIUM, RUTHENIUM, OSMIUM,** in the word search below .

R	E	B	H	C	A	R	U	Y	T
M	U	I	D	A	L	L	A	P	S
P	V	T	E	C	F	H	I	R	X
I	L	B	H	O	X	A	L	H	E
O	Y	A	B	E	J	D	L	O	G
R	S	T	T	V	N	O	P	D	A
E	A	M	U	I	D	I	R	I	S
W	O	K	I	J	N	H	U	U	F
C	O	N	E	U	D	U	S	M	T
R	P	A	E	W	M	I	M	L	A

When you have found them, look them up in the Periodic Table. Draw a block of these elements as they appear in the Periodic Table.

a)   These metals are often found with nickel. Use the Periodic Table to predict which other metal or metals might be present. Add these to your blocks.

b)   Gold is also found in the ore. Name another precious metal that you might expect to be present.

The solvent extraction of platinum

The last stage of extraction of platinum involves dissolving the magnetic concentrate produced from the matte. This concentrate is the solid obtained after smelting and magnetic separation. After removal of the base metals, the concentrate is dissolved in hydrochloric acid to give a solution of the platinum as a complex ion, $[PtCl_6]^{2-}$(aq), the hexachloroplatinate(IV) ion, along with other complex ions.

Magnetic concentrate from processed matte

+

6 mol dm^{-3} hydrochloric acid

↓

$[PtCl_6]^{2-}$ (aq) + other complex ions

Once the platinum is in solution it can be removed selectively by exchange of ions using amines dissolved in organic solvents. In this process, the metals are extracted using the technique of partition.

The partition coefficient, K, is :

$$K = \frac{[\text{Concentration of solute in organic layer}]}{[\text{Concentration of solute in aqueous layer}]}$$

so if K is greater than 1, there is a greater proportion of solute in the organic layer.

1. Demonstration

   Iodine dissolves in both the aqueous solvents and the organic solvents:

   I_2(aq) $\rightleftharpoons$ I_2(org)

   a) Which layer is the organic layer?

   b) (i) In which layer is the iodine most soluble?

   (ii) Explain why you might expect this.

   (iii) Would K for this example be greater than, less than or equal to 1?

2. The organic layer used in the solvent extraction of platinum could be a solution of primary, secondary, tertiary amines or quaternary ammonium salts.

   Label the following amines as primary, secondary, tertiary and quaternary cations.

   a) R—N: (with R above and R below)   b) R—N: (with R above and H below)   c) [R—N—R]$^+$ (with R above, R left, R right, R below)   d) R—N: (with H above and H below)

3. Predict the equation for the reaction between a primary amine and hydrochloric acid.

4. The figure shows the distribution data for the extraction of $[PtCl_6]^{2-}$ from an aqueous solution by an organic solvent containing different amines.

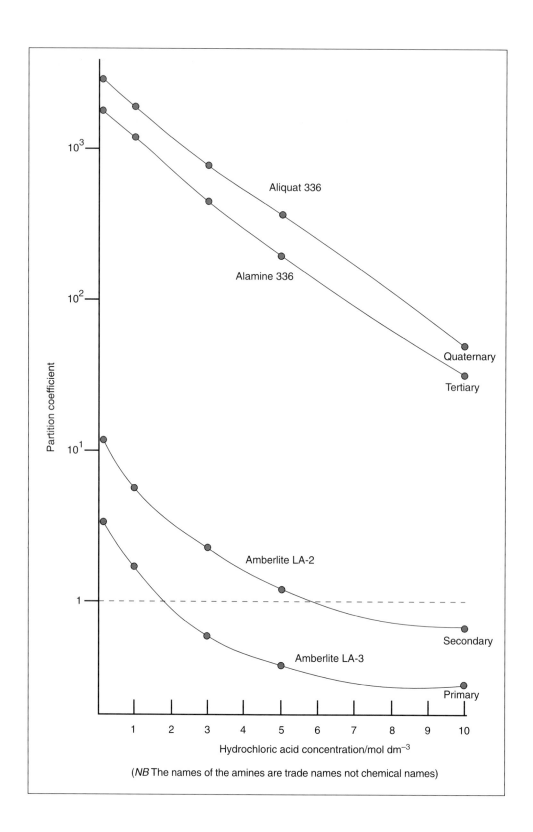

(*NB* The names of the amines are trade names not chemical names)

a)   Which amine is the most efficient for extracting $[PtCl_6]^{2-}$?

b)   What is the significance of the dotted line at partition coefficient value 10^0?

c)   At 7 mol dm^{-3} hydrochloric acid, for which amine(s) does the $[PtCl_6]^{2-}$ dissolve more in the aqueous acid than in the organic amine-containing layer?

Aspects of the platinum group metals (PGM)

The PGM iridium, osmium, palladium, platinum, rhodium, and ruthenium are commonly found together in ores, along with silver, gold and base metals. To understand how they are separated from each other and purified, it is necessary to to know something of their chemistry.

The PGM are found in the second and third row of the transition metals in the Periodic Table, see below.

d-block

45.0 Sc Scandium 21	47.9 Ti Titanium 22	50.9 V Vanadium 23	52.0 Cr Chromium 24	54.9 Mn Manganese 25	55.8 Fe Iron 26	58.9 Co Cobalt 27	58.7 Ni Nickel 28	63.5 Cu Copper 29	65.4 Zn Zinc 30
88.9 Y Yttrium 39	91.2 Zr Zirconium 40	92.9 Nb Niobium 41	95.9 Mo Molybdenum 42	(99) Tc Technetium 43	101.1 Ru Ruthenium 44	102.9 Rh Rhodium 45	106.4 Pd Palladium 46	107.9 Ag Silver 47	112.4 Cd Cadmium 48
138.9 La Lanthanum 57	178.5 Hf Hafnium 72	181.0 Ta Tantalum 73	183.9 W Tungsten 74	186.2 Re Rhenium 75	190.2 Os Osmium 76	192.2 Ir Iridium 77	195.1 Pt Platinum 78	197.0 Au Gold 79	200.6 Hg Mercury 80

The PGM plus gold and silver make up the Precious Metals.

They have very similar chemical behaviour which makes separation difficult. However, like the other d-block elements the PGM are able to form complexes with a wide variety of ligands. From a practical view the most important are the chloro-complexes, *eg* $[PtCl_6]^{2-}$, because in the final stage of processing the ore, the PGM concentrate is treated with 6 mol dm^{-3} hydrochloric acid to dissolve all the precious metals. The result is a mixture of complexes in which the elements can have several different oxidation states. These are summarised overleaf.

Ruthenium		Rhodium		Palladium		Silver	
Ru(III)	$[RuCl_6]^{3-}$ $[RuCl_5(H_2O)]^{2-}$ $[RuCl_4(H_2O)_2]^-$ $[RuCl_3(H_2O)_3]$	Rh(III)	$[RhCl_6]^{3-}$ $[RhCl_5(H_2O)]^{2-}$ $[RhCl_4(H_2O)_2]^-$	Pd(II)	$[PdCl_4]^{2-}$	Ag(I)	AgCl
Ru(IV)	$[RuCl_6]^{2-}$ $[Ru_2OCl_{10}]^{4-}$ $[Ru_2OCl_8(H_2O)_2]^{2-}$			Pd(IV)	$[PdCl_6]^{2-}$		
Osmium		**Iridium**		**Platinum**		**Gold**	
Os(III)	$[OsCl_6]^{3-}$ $[OsCl_5(H_2O)]^{2-}$ $[OsCl_4(H_2O)_2]^-$	Ir(III)	$[IrCl_6]^{3-}$ $[IrCl_5(H_2O)]^{2-}$ $[IrCl_4(H_2O)_2]^-$	Pt(II)	$[PtCl_4]^{2-}$	Au(III)	$[AuCl_4]^-$
Os(IV)	$[OsCl_6]^{2-}$	Ir(IV)	$[IrCl_6]^{2-}$	Pt(IV)	$[PtCl_6]^{2-}$		

The precious metals and their oxidation states in hydrochloric acid

a) Apart from gold, the elements form a series of aquated chloro-complexes when in oxidation state +III.

b) Ruthenium is unique among the PGM in that it also forms a series of oxo-bridged dimers in oxidation state +IV.

c) The chloro-aquocomplexes of an element exist in equilibrium with each other. For example, the equilibrium distribution diagram for rhodium(III) shows how the relative concentrations of the complexes change with increasing concentration of the chloride ion.

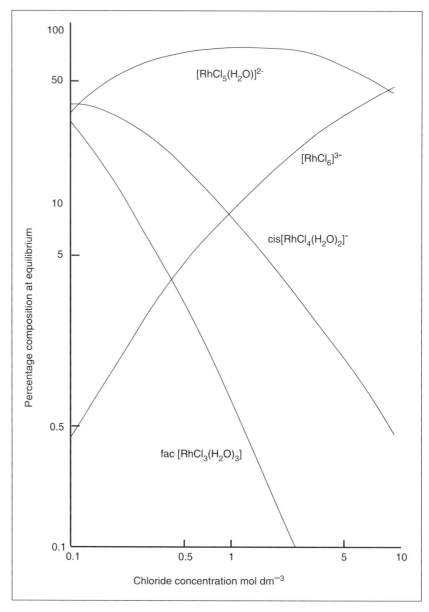

**Equilibrium distribution diagram for Rh(III)
(note the non-linear scales on the axes)**

This has important practical implications, because the PGM present in concentrated hydrochloric acid are separated from each other on the basis of the different solubilities of their complexes in organic solvents. In general the following order of solubility is seen in the organic phase:

$$[MCl_4]^- > [MCl_6]^{2-} > [MCl_6]^{3-}$$

This is thought to be for steric reasons. It is harder to pack three bulky organic monocations around $[MCl_6]^{3-}$ than it is to place two around $[MCl_6]^{2-}$ or one around $[MCl_4]^-$.

PGM chloro-complexes can be extracted in one of two ways.

1   Ligand exchange

2   Anion exchange

Ligand exchange

The PGM are generally much less reactive than the base metals. With base metal complexes one ligand can easily be substituted for another, eg

$$[Cu(H_2O)_6]^{2+}(aq) + 4Cl^-(aq) \rightarrow [Cu\,Cl_4]^{2-}(aq) + 6H_2O(l)$$

This reaction occurs rapidly, but equivalent reactions amongst the PGM can take hours, days or even months under standard conditions. The kinetics are summarised below.

Ruthenium	Rhodium	Palladium	Silver
Ru(III) 10^{-3}–10^{-4} Ru(IV) 10^{-5}–10^{-6}	Rh(III) 10^{-3}–10^{-4}	Pd(II) 1	Ag(I) 10^4–10^6
Osmium	Iridium	Platinum	Gold
Os(III) 10^{-7}–10^{-9} Os(IV) 10^{-10}–10^{-12}	Ir(III) 10^{-4}–10^{-6} Ir(IV) 10^{-8}–10^{-10}	Pt(II) 10^{-3}–10^{-5} Pt(IV) 10^{-10}–10^{-12}	Au(III) 10^1–10^{-1}

Relative substitution kinetics of the precious metal chloro-complexes (palladium = 1) Note: The smaller the substitution constant the more stable the complex

There is a wide variation between the different PGM and their oxidation states. Silver(I) has a reactivity approaching that of the base metals, gold(III) is relatively reactive but osmium(IV) is inert. In practice only palladium(II) is extracted into the organic layer via ligand substitution as gold has been removed previously.

Anion exchange

The remaining PGMs are mostly extracted by anion exchange in which their complexes replace chloride ions, eg

$$2[R_2NH_2^+\,Cl^-] + [PtCl_6]^{2-}(aq) \rightleftharpoons (R_2NH_2^+)_2\,[PtCl_6]^{2-} + 2Cl^- \quad (R_2NH = amine)$$

The PGM, like all transition metals, have variable oxidation numbers and therefore can participate in redox reactions. For example, during the separation process, iridium(III) is first oxidised to iridium(IV), extracted, then reduced back again in the organic phase. The redox behaviour of the PGM in acidic chloride media is summarised overleaf.

Ruthenium	Rhodium	Palladium	Silver
Ru(III) $[RuCl_5(H_2O)]^{2-}$	Rh(III) $[RhCl_6]^{3-}$	Pd(II) $[PdCl_4]^{2-}$	Ag(I) AgCl
⇕ 0.83 V	⇕ >1.4 V	⇕ 1.29 V	
Ru(IV) $[Ru_2OCl_{10}]^{4-}$	Rh(IV) $[RhCl_6]^{2-}$	Pd(IV) $[PdCl_6]^{2-}$	
⇕ >1.4 V			
Ru(VIII) RuO_4			
Osmium	Iridium	Platinum	Gold
Os(III) $[OsCl_6]^{3-}$	Ir(III) $[IrCl_6]^{3-}$	Pt(II) $[PtCl_4]^{2-}$	Au(I)* $[AuCl_2]^{-}$
⇕ 0.42 V	⇕ 0.96 V	⇕ 0.74 V	⇕ 0.93 V
Os(IV) $[OsCl_6]^{2-}$	Ir(IV) $[IrCl_6]^{2-}$	Pt(IV) $[PtCl_6]^{2-}$	Au(III) $[AuCl_4]^{-}$
⇕ 1.0 V			
Os(VIII) OsO_4			

* Au(I) slowly disproportionates $3[AuCl_2]^- \longrightarrow [AuCl_4]^- + 2Au + 2Cl^-$

Redox behaviour of precious metals in acidic chloride media

Note

The more positive E^θ, the more oxidising the conditions must be for a reaction to take place.

The stability of the higher oxidation states tends to decrease from left to right across the precious metals. For example, silver usually only forms stable complexes with an oxidation number of +I, while ruthenium and osmium form stable tetroxides with an oxidation number of +VIII under highly oxidising conditions. This enables these elements to be separated from the other PGM, as their tetroxides are volatile.

Questions

1. Explain the following terms, giving examples from the text where possible.

    a) Ligand.

    b) Complex.

    c) Coordination number.

    d) Oxo-bridged dimer.

    Suggest a structure for an oxo-bridged dimer.

2. From the figure listing the precious metals and their oxidation states in hydrochloric acid, give an example of a complex in which a PGM has the oxidation number:

    a) +II

    b) +III

    c) +IV

3. a) Predict the shapes of the following complexes:

    (i) $[PtCl_6]^{2-}$

    (ii) $[Ru_2OCl_{10}]^{4-}$

    b) What isomers are possible for the following complexes (a set of molecular models might help): $[PtCl_2(H_2O)_2]$ (this is square planar), $[RuCl_4(H_2O)_2]^-$, $[RuCl_3(H_2O)_3]$

4. The graph shows the equilibrium distribution of rhodium(III) chlor-aquo-species at 25 °C. List the complexes in order of relative abundance at

    a) $[Cl^-] = 0.3$ mol dm^{-3}

    b) $[Cl^-] = 5$ mol dm^{-3}

5. For each pair of complexes, say which you think is more soluble in organic solvents compared with the aqueous phase. Explain your reasons for the answer you choose.

    a) $[MCl_6]^{2-}$ or $[MCl_4]^-$

    b) $[MCl_6]^{2-}$ or $[MCl_5(H_2O)]^{2-}$

This page has been intentionally left blank.

RS•C

Smart materials

Contents

RS•C

Introduction

Plastics show an ever increasing number of properties which makes them an exciting part of materials science. This pack introduces **electrically conducting plastics** and **heat-shrinking plastics.** The concepts in this pack are relevant to both pre-16 and post-16 chemistry students. The material allows revision of basic chemistry via interesting and up-to-date material.

This pack consists of:

▼   Teacher's notes for each part, which include the answers to the exercises;

▼   two comprehension exercises for pre-16 students:

Conducting plastics
This introduces plastics which conduct electricity, their properties and uses; and

Shape changing polymers (or molecules with a memory)
This covers shape changing polymers, properties and uses; and

▼   two comprehension exercises for post-16 students:

Conducting polymers
This includes information on intrinsically and extrinsically conducting polymers; and

Shape memory polymers
The chemistry involved in crosslinked polymer formation, both by irradiation and by chemical methods, is considered.

This pack is the result of a Royal Society of Chemistry Study visit to Raychem Ltd in December 1995 and was based on original ideas produced by:

▼   Penelope Bagshaw, Langley Grammar School;

▼   David Cooper, Sutton Valence School;

▼   Paul Evans, Hethersett High School;

▼   Jean Johnson, consultant;

▼   Chris Law, West Thames College; and

▼   Alan Tin-Win, The Beaconsfield School.

THE ROYAL
SOCIETY OF
CHEMISTRY

RS•C

Part 1 Pre-16 – Conducting plastics and shape changing polymers

Teacher's notes

These worksheets consist of comprehension exercises that revise many basic ideas from chemistry and approach them from a novel perspective. They could be used for homework or self-study.

Answers to questions on conducting plastics

1. A polymer is a large molecule made up of many linked repeating units (monomers) of smaller molecules.

2. Any suitable use – *eg* electric plugs, sheathing for wire, bodies for hair driers, handles for kettles, irons.

3. Any metal or graphite (most forms of carbon conduct to some extent).

4. More carbon could be added to improve its conductivity.

5. It could melt or catch fire if overheated.

6. a) A liquid becomes more viscous as it cools.

    b) A typical liquid freezes.

    c) Water pipes will crack if the water freezes. Pipes carrying a viscous liquid – *eg* chocolate, sulfur, oil – may block if the liquid cools.

    d) Coil a self-regulating heating wire round the pipe to keep the contents warm.

7. A polymer – which is an insulator.

    Carbon – which conducts electricity.

8. Only the parts of the pipe that were cold would be heated, therefore it is more efficient and cheaper to run. The cable cannot overheat.

9. The particles have more energy and therefore vibrate (solid) or move (liquids and gases) further apart from each other. The material thus expands.

10. The material consists of thousands of independent parallel circuits each of which responds to its ambient temperature.

RS•C

Answers to questions on shape changing polymers (or molecules with a memory)

1. Thermosoftening plastics can be recycled because these plastics can be melted down and remoulded.

2. To *melt easily* means to turn into a liquid at a low temperature.

3. A rigid, brittle material is hard to bend and breaks rather than bends.

4. Typical properties of plastics:

   ▼ poor conductor;

   ▼ water resistant; and

   ▼ low density.

5. A loose sleeve is fitted round the wires and then heated to make it shrink to form a tight seal.

6. Any suitable ideas – *eg* to protect junctions in cables for phones, electricity.

7. a) It softens when heated and its shape can be changed.

   b) It has a fundamental shape.

8. This is the property of returning to a given shape when heated.

RS•C

Part 2 Post-16 – Conducting polymers and shape memory polymers

Teacher's notes

These are worksheets for post-16 chemistry students.

The section *Conducting polymers* uses the new material as a basis for comparing the structures of intrinsically and extrinsically conducting polymers.

The section *Shape changing polymers* is suitable for use with students who have already studied simple hydrocarbons, and addition polymerisation, and have been introduced to free radical mechanisms, including that for the reaction of methane with bromine. It uses the concepts of thermosetting and thermosoftening plastics to consider polymer formation by both irradiation and chemical reactions.

Answers to questions on conducting polymers

1. Having alternating single and double bonds.

2.

   a)   all *trans-* poly(ethyne)

   b)   all *cis-* poly(ethyne)

3. An alkene, *not* an alkyne.

4. Because oxygen attacks the electron-rich double bonds in alkenes. Sensible suggestions for products include: epoxy compounds, ozonides, diols and ketones (with the chain breaking).

5. Suggestions might include;

   low density; and

   easy to shape at lower temperatures.

6.

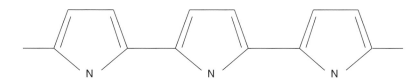

7. A material composed of a mixture of two or more materials. The composite's properties are derived from those of its constituents.

RS•C

8. The conductivity goes down with increasing temperature – *ie* the resistance goes up.

9. Structure a) as it is conjugated.

10. C–C, 0.134 nm; C=C, 0.154 nm; benzene, 0.140 nm. X-ray diffraction to measure the bond lengths.

11. Electrons move through a "sea" (the conduction band) which encompasses the whole metal. This sea is three-dimensional rather than two-dimensional as in graphite and one-dimensional as in poly(ethyne) and can be thought of as a super-delocalised orbital.

Answers to questions on shape memory polymers

1. C–C bonds. This could degrade the polymer by shortening the chains.

2. The polymer might become, in effect, a thermoset. This means it is no longer mouldable.

3. Van der Waals forces. They are typically 1/100 the strength of a covalent bond – a few kJ mol^{-1} compared with about 350 kJ mol^{-1}. The high strength of the plastic is because large numbers of these bonds must be broken to break the material.

4. a) Propagation and termination.

    b) Ultraviolet light does not have enough energy to break C–H bonds. (Typically ultraviolet light has an energy of 400 kJ mol^{-1} compared with 413 kJ mol^{-1} for an average C–H bond). Ultraviolet light has less energy per quantum than β-radiation.

    c) $CH_3^{\bullet} + CH_3^{\bullet} \rightarrow C_2H_6$ as a C–C bond is formed in both cases.

5. *Homolytically* describes the breaking of a covalent bond so that one of the shared electrons goes to each of the atoms in the bond.

    Electronegativity is the ability of an atom to attract electrons to itself in a covalent bond.

Conducting plastics information sheet

Usually plastics are excellent electrical insulators. They do not conduct electricity. However, they can be made to conduct electricity by mixing them with a material that is a good conductor. This material is called a **conducting filler**. The plastic polymer holds the filler in position so that it carries the electric current through the polymer. Changing the ratio of filler to polymer changes how well the material conducts. Many commercial products use carbon black – a form of carbon that conducts electricity – as the conducting filler.

Self-regulating heating cables

An electric current always has a heating effect. So, like all electrical components, conducting polymers warm up when an electric current flows through them. These polymers are used to make heaters with a built in temperature controller. The heaters are in the form of cables that, for example, can be attached to pipes to keep their contents at a particular temperature.

How they work

Two copper wire electrodes run along the whole length of cable. They are held apart, embedded in a conductive polymer which contains carbon black. The carbon black particles form tiny conducting paths between the two electrodes. This results in thousands of parallel circuits along the length of the cable, all acting independently of each other.

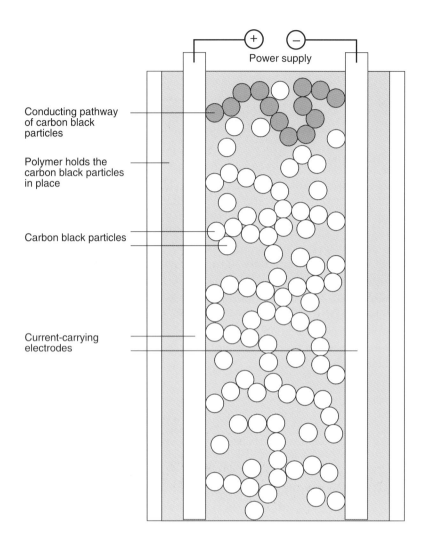

One complete conducting pathway of carbon black particles has been shaded in. Can you spot four others?

If we consider a section of the cable, we can see how it works (see below).

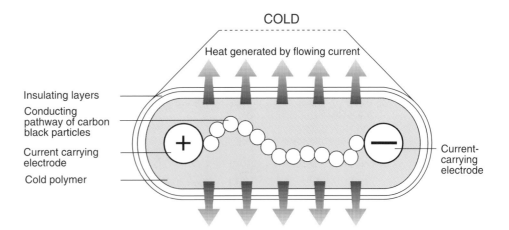

COLD

Heat generated by flowing current

Insulating layers

Conducting pathway of carbon black particles

Current carrying electrode

Cold polymer

Current-carrying electrode

As electric current flows through the carbon black, heat is generated. This is the source of heat in the cable and this heat may be used, for example, to keep the contents of a pipe at the correct temperature. But, as the polymer itself warms up, it expands. As it expands, the carbon black particles inside it are pulled apart (see below). This breaks some of the conducting paths, less current flows, less heat is generated and the cable cools down.

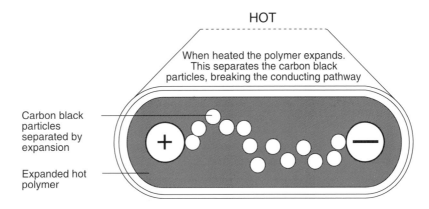

HOT

When heated the polymer expands. This separates the carbon black particles, breaking the conducting pathway

Carbon black particles separated by expansion

Expanded hot polymer

As it cools, the polymer contracts, forcing the carbon black particles together again and reforming the conductive paths. The cable generates more heat again.

The self-regulating heating cable has many advantages over conventional electrical heaters:

▼ heat is only generated when required. This makes the heater energy efficient;

▼ the heat generated varies as the surrounding temperature varies;

▼ heat is only generated in the parts of the cable which are cold; the heating on the cable is localised. This is because the conductive carbon black pathways act independently as thousands of miniature parallel resistors;

▼ the cable is flexible, allowing the heating of awkward shapes such as pipes;

▼ as it is manufactured from a polymer, the cable is corrosion resistant;

▼ by altering the electrical properties of the polymer a variety of temperatures may be achieved making it suitable for a large range of applications;

▼   the cable can be cut to the required length up to a maximum of 250 m; and

▼   the cable cannot overheat, even when overlapped or sandwiched in thermal insulation, thus increasing safety.

Questions

1.   What is a polymer?

2.   Give two examples of where a plastic would be used because of its electrically insulating properties.

3.   Give an example of a material that conducts electricity.

4.   How could the conducting plastic be manufactured so that it becomes a better conductor of electricity?

5.   What might happen to the plastic polymer if it overheats?

6.   a)   What happens to the runniness (viscosity) of a liquid such as treacle when it cools?

     b)   What happens to a typical liquid if it is cooled to a low temperature?

     c)   Give two examples of situations where it is important that the contents of a pipe do not cool down too much. Explain what might happen if they did.

     d)   How could you use conducting polymers to keep the contents of a pipe warm?

Extension questions

7.   Self-regulating heating cable is made of a **composite** material. This is a mixture of two or more materials and combines the properties of the materials in it. What two main materials are present in this composite and what property of each is necessary?

8.   List any advantages in using a self-regulating heating cable to keep the contents of a pipe at the correct temperature, compared with using a conventional electric heater.

9.   Expain in terms of the behaviour of their particles why materials expand when they are heated.

10. In self-regulating heating cables, *heat is only generated when and where it is required.* Explain why this is. You may wish to use diagrams in your answer.

Shape changing polymers (or molecules with a memory) information sheet

One way to classify polymers (plastics) is by their response to heat. All plastics can be moulded when they are made, and then set hard into shape as they cool. After this, **thermosoftening** (also called thermoplastic) plastics can be remelted, and then reshaped. **Thermosetting** plastics cannot be remelted and so cannot be reshaped once they have been made.

If we want a flexible plastic carrier bag, then a thermosoftening plastic is suitable. On the other hand, a saucepan handle must be made of a thermosetting plastic. This can withstand quite high temperatures. At very high temperatures it eventually burns or decomposes rather than melts.

There is now a type of polymer which makes a plastic that is midway between these two sorts. It has a very special property called **shape retention**. When this plastic has been heated it softens and becomes pliable. In this state it can be stretched. If it is cooled it will set hard in this stretched shape. But, the plastic remembers its original shape and will, if heated again, shrink back to this. It is called a **heat-shrinking polymer (or plastic)**.

A question of structure

Thermosoftening plastics can be thought of as long chains of large molecules which have few or no crosslinking covalent bonds between the chains (see below). This sort of structure is easily melted as the chains will slide away from each other. If the material is stretched when it is hot the chains slide past each other and then stay in their new position, when the material is cooled.

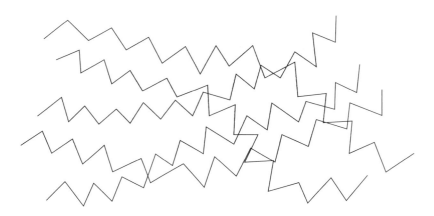

A thermosoftening plastic

Thermosetting plastics have many covalent bonds linking the long chains. Not only is it impossible to melt the polymer if the material is stretched, the chains cannot slide past one another. The material is rigid and brittle.

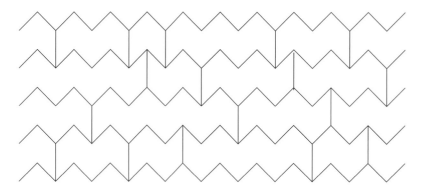

A thermosetting plastic

Heat-shrinking plastics have some crosslinking between the chains – more than thermosoftening plastics and fewer than thermosetting plastics.

They can be heated and softened enough to be reshaped. This stretches the cross-links. When they are cooled in the stretched state they stay stretched and retain the new shape. But, if they are now heated the chains are free to slide back to their original position, and the plastic returns to its original shape. The bonds are then behaving rather like stretched elastic bands that have been released.

Why are heat-shrinking plastics useful?

They can be used in awkward situations because when they shrink they can fit tightly over an object. A common example of this is heat shrink plastic sleeving for electrical wiring, for instance under the bonnet of a car. It is easy to fit a loose sleeve round a join and then seal it tightly by applying heat.

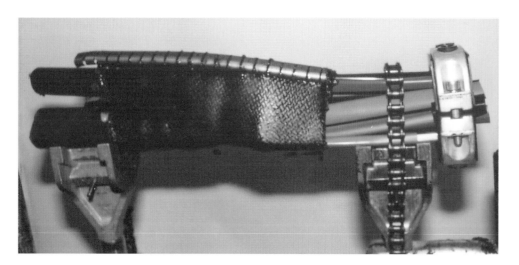

Heat shrink sleeving before being heated

Questions

1. Which sort of plastic, thermosoftening or themosetting, is suitable for recycling? Explain your answer.

2. What does *to melt easily* mean?

3. What are the properties of a *rigid, brittle* material?

4. From the list below pick out the properties of a typical plastic:

   good conductor of electricity/poor conductor of electricity

   water resistant/water absorbent

   high density/low density

5. Explain how a heat-shrinking plastic could be used to hold together a bunch of wires in a car engine.

6. Where else could a heat-shrinking plastic be used to provide a tight waterproof seal?

Extension questions

7. In what way does a heat-shrinkable plastic resemble:

   a) a thermosoftening plastic?

   b) a thermosetting plastic?

8. What is meant by *shape retention*?

Conducting polymers information sheet

Most textbooks indicate that one of the most important properties of polymers is that they are electrical insulators – they are used for covering electrical cables, the bodies of electrical plugs and sockets, and so on. This is no longer completely true. Over the past few years several polymeric materials have been produced that conduct electricity and a range of applications is being developed. These conducting polymers are of two basic types:

▼ **intrinsically conducting** polymers where the polymeric material itself conducts; and

▼ **extrinsically conducting** polymers which are composites where a conductive material such as carbon black is embedded in a non-conducting polymer such as poly(ethene).

Intrinsically conducting polymers

The simplest intrinsically conducting polymer is poly(ethyne), sometimes called poly(acetylene), (see below) which, despite its name, is an alkene not an alkyne. It consists of a hydrocarbon chain with alternating single and double bonds; called a **conjugated** system. The p-orbitals which form the double bonds can overlap to form a delocalised π–system (similar to the one in benzene). Electrons flow through the delocalised system and so the polymer can conduct. In fact, additives such as iodine have to be incorporated to maximise the conductivity by ensuring that the polymer does exist in the delocalised form rather than as localised single and double bonds. Suitably doped poly(ethyne) can have a conductivity comparable with that of copper provided the material has been stretched to align the chains so that they all run in the same direction. Poly(ethyne) has problems for everyday applications as it is attacked by oxygen from the air but other more stable polymers with conjugated systems also have conducting properties. There are some examples on the next page.

$$H - C \equiv C - H \qquad H - C \equiv C - H \qquad H - C \equiv C - H \qquad \text{ethyne}$$

↓

poly(ethyne)

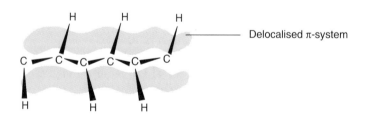

Poly(ethyne)

Delocalised π-system

Poly(ethyne)

Poly(alkylthiophene)

Poly(aniline)

Poly(thiophene)

Poly(pyrrole)

Poly(furan)

Examples of intrinsically conducting polymers

The well-known conductivity of graphite (see below) can be explained in the same way. Here there is a two-dimensional delocalised system covering a layer of carbon atoms so that graphite conducts well along the planes of carbon atoms but poorly at right angles to them.

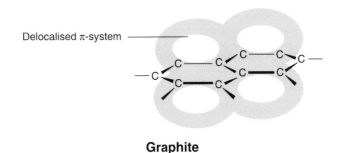

Graphite

Extrinsically conducting polymers

One type of extrinsically conducting polymer consists of a matrix of poly(ethene) with a percentage of conducting carbon black (a form of powdered graphite) incorporated in it. If the carbon black particles are close enough to be in contact with one another, the material conducts. If the particles are not in contact, it is an insulator. This means that the degree of electrical conduction depends on temperature. At high temperature, the poly(ethene) matrix expands and pulls the particles of carbon black away from each other, decreasing the conductivity. At lower temperatures the poly(ethene) contracts, the carbon black particles are closer and the material conducts well. This temperature dependence of conductivity leads to the use of this material in self-regulating heater cable and PolySwitch* re-settable circuit protection devices.

* PolySwitch is a registered trademark of Raychem Corporation.

Beat the freeze – the IceStop‡ system

Ice can cause a lot of damage – burst pipes, slippery walkways, collapsing roofs - all of which can be prevented by low level heating. Conventional heating circuits have some disadvantages here as they have a constant current which can result in 'hot spots' and energy wastage. The IceStop system consists of parallel copper wires embedded in a conducting polymer. Carbon granules form conducting pathways between the wires resulting in a large number of miniature parallel circuits. The polymer conducts electricity well and thus acts as a heater, only when it is cold. As the material warms up the poly(ethene) expands, interrupting some of the conducting pathways and switching off the miniature circuits (see below).

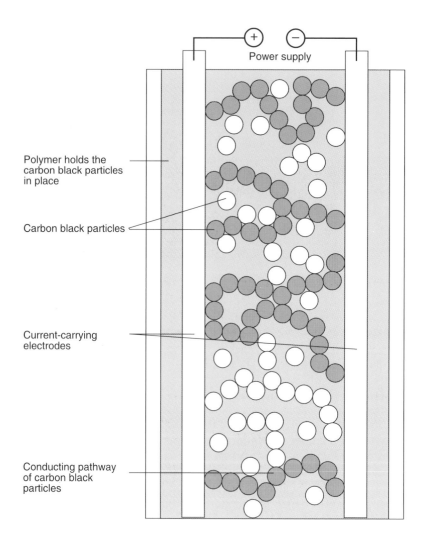

Heat is generated only where electric current flows through the carbon black pathways

‡IceStop is a registered trademark of Raychem Corporation.

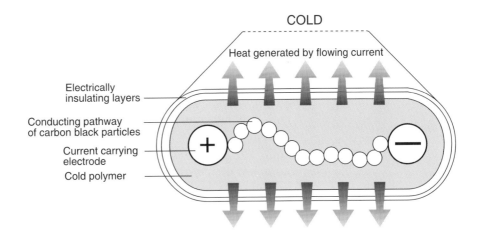

COLD

Heat generated by flowing current

Electrically insulating layers

Conducting pathway of carbon black particles

Current carrying electrode

Cold polymer

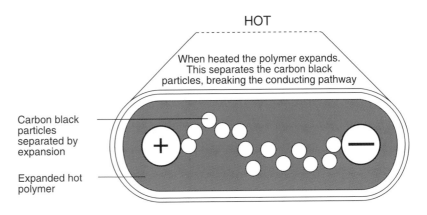

HOT

When heated the polymer expands. This separates the carbon black particles, breaking the conducting pathway

Carbon black particles separated by expansion

Expanded hot polymer

Heater cable

IceStop cable can be laid along water pipes, under pathways and along guttering to provide low-level, self-regulating heating which keeps the environment frost-free. It is flexible and easy to install, can be cut to any length required and can be overlapped or wound round a pipe.

Protecting batteries – the PolySwitch device

Lithium batteries are used in many small hand-held electrical appliances such as cameras but a current overload can lead to overheating resulting in, at best, damage to the appliance, and, at worst, the battery exploding. Lithium batteries are also used in telecom systems, audio speakers, fire and burglar alarms and personal computers. A conventional fuse could provide the required protection but needs to be replaced if it blows. A PolySwitch device does the same job but can be reset, rather than having to be replaced, once the fault has been rectified. As in IceStop cable, carbon granules form conducting pathways through the polymer and these pathways are broken if the material becomes too warm. This protects the appliance from current overload.

> **Why doesn't the fuse keep resetting itself?**
> When the PolySwitch device gets hot, it does not switch off the current completely as does a melted wire in a fuse. A very small current still flows through the device. This is enough to keep it hot. Once the fault has been rectified, the PolySwitch device can be reset by first turning off the power to allow it to cool.

Questions

1.   Explain the term conjugated.

2.   The groups attached to the double bonds in poly(ethyne) can be either *cis* or *trans*. Poly(ethyne) exists in two extreme forms – all *cis–* and all *trans–*. Draw the structure of each. You should draw at least five repeating units.

3.   What is the functional group in poly(ethyne)?

4.   Why would you expect oxygen to attack poly(ethyne)? Suggest a possible product of the reaction.

5.   Suggest some advantages which a conducting polymer might have over a conducting metal. Assume that the typical polymer properties are essentially unchanged (except for electrical conductivity).

6.   Draw a displayed formula for poly(pyrrole) showing at least three repeating units.

7.   What is a composite material? What affects its properties?

8.   Does the electrical conduction of metals rise or fall as the temperature increases? (Hint: think about superconductivity.)

9.   Look at the formulae of the two polymers below. Predict which one you might expect to be an intrinsic conductor and explain your choice.

a)

b)

Conducting polymers

10.   Use a data book to find the bond lengths you would expect for C–C and for C=C. What is the carbon–carbon distance in benzene? What technique might you use to find out if poly(ethyne) were in the delocalised or non-delocalised form?

11.   Describe the way in which electrons move in metals and allow them to conduct electricity. Compare this with the situation in poly(ethyne) and graphite. Write down any similarities and differences.

Shape memory polymers information sheet

These polymers 'remember' the shape into which they have been moulded and will return to it on gentle heating.

They are based on thermoplastic polymers. During manufacture the polymer is moulded into a particular shape and irradiated with β-radiation. It is then heated, re-shaped and cooled. It does, however, remember the shape which it had when irradiated and returns to it when re-heated. One application of this effect is heat-shrinkable sleeves which are used to hold together bundles of wires in car wiring harnesses (see below).

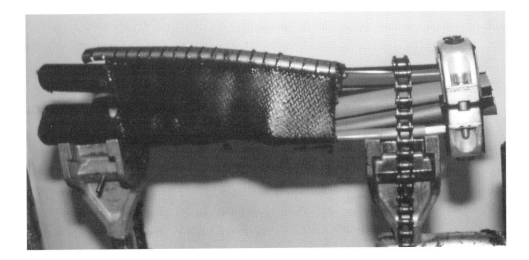

Applying a shrinkable sleeve

Here a cylindrical length of poly(ethene) sleeve is moulded into shape so that it has a narrow internal diameter which firmly holds together a bunch of wires. The sleeve is now irradiated, which causes covalent crosslinks between the poly(ethene) chains.

The tube is next heated to above its crystalline melting point to soften it. Then it is stretched into a larger diameter. This stretches the crosslinks. The tube is cooled and this locks the chains in their stretched position.

Now the large diameter tube can easily be slipped over a bunch of wires. If it is heated (by a hot air gun or blowlamp) above its crystalline melting point, the stretched crosslinks pull the material back into the shape it had on irradiation and it holds the bunch of wires firmly together.

Bonding within polymers

Polymers can be classified as thermoplastic (thermosoftening) or thermosetting. Thermoplastics soften on heating and can be moulded into a shape which they retain on cooling. They can be reheated and moulded indefinitely. They consist of long chain molecules, each chain being essentially independent of the others. There are no covalent bonds between the chains. The plastics retain their shape when cool because of intermolecular forces between the chains. In particular there are areas where the the chains line up in an ordered way – so-called areas of crystallinity. If crystalline areas on two adjacent chains line up, the intermolecular interactions are particularly strong. This is responsible for much of the strength of thermoplastics in the solid state. On heating above the crystalline melting point, increased thermal motion makes the crystalline areas disappear. The polymer softens, the chains become free to move past one another and the plastic can be moulded. On cooling, new crystalline areas re-form, which help retain the new shape (see below).

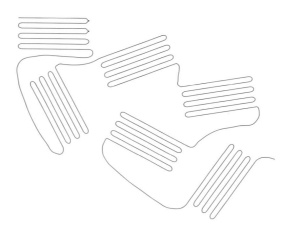

Crystalline areas in a polymer chain

Thermosetting plastics have many covalent crosslinks between the polymer chains which form as the polymer is made. Once made, the polymer is unaffected by heat (until it begins to burn or decompose).

Shape memory plastics have a degree of crosslinking (after irradiation) which is less than that of a thermoset but more than that of a thermosoftening plastic.

Irradiation

How does the irradiation process produce the cross links? β-radiation is a stream of electrons each with more than enough energy to break covalent bonds. β-irradiation of poly(ethene) breaks some of the C–H bonds in the poly(ethene) chains. As carbon and hydrogen atoms have similar electronegativity, the bonds tend to break homolytically leaving a free hydrogen atom and a carbon free radical, a carbon atom with a single –*ie* unpaired – electron. Such carbon atoms are extremely reactive and two close together may form a covalent bond thus pairing up their electrons. This forms a crosslink between the chains (see below).

The effect of β-irradiation on poly(ethene)

The hydrogen atoms, which also have an unpaired electron each, tend to come together to form hydrogen molecules which escape from the polymer.

Questions

1. What other bonds will the β-radiation break? Suggest what effects this might have on the polymer.

2. What might happen if too many crosslinks are formed in the polymer? What effect might this have on its properties?

3. What are the intermolecular forces which operate within the crystalline regions of a thermosoftening plastic called? Compare the strength of a single intermolecular interaction with that of a typical covalent bond. Explain why, when cool, these intermolecular forces have comparable effects to covalent bonds – *ie* thermosets and thermoplastics have comparable strengths.

4. Another free radical reaction is that of bromine with methane in ultraviolet light. The steps are:

   1. $Br–Br \longrightarrow 2Br^{\bullet}$

   2. $CH_4 + Br^{\bullet} \rightarrow HBr + CH_3^{\bullet}$

      and

      $Br_2 + CH_3^{\bullet} \rightarrow Br^{\bullet} + CH_3Br$

   3. $Br^{\bullet} + Br^{\bullet} \xrightarrow{UV} Br_2$

      $CH_3^{\bullet} + CH_3^{\bullet} \rightarrow C_2H_6$

      $Br^{\bullet} + CH_3^{\bullet} \rightarrow CH_3Br$

   a) The first step is called initiation. Name the other two.

   b) In this case, there are no hydrogen free radicals formed by the reaction

      $CH_4 \rightarrow H^{\bullet} + CH_3^{\bullet}$

      Suggest a reason for this.

      What does this tell you about the energy of ultraviolet light compared with that of β-radiation?

   c) Which of the three possibilities for step 3 above is most similar to the reaction which occurs in the polymer crosslinking reaction? Explain your choice.

5. Explain the terms *homolytically* and *electronegativity* as used in the passage.

This page has been intentionally left blank.

Biodiesel

Contents

RS•C

THE ROYAL
SOCIETY OF
CHEMISTRY

Introduction

This pack is about the fuel biodiesel, which has similar properties to diesel obtained from crude oil, but is manufactured from vegetable oils and, in particular, rape seed oil. Biodiesel has many environmental advantages over fossil diesel as well as being a renewable fuel. The aim of this pack is to provide interesting and relevant material for both pre-16 science/chemistry students and post-16 chemistry students.

The pack includes:

▼ instructions for practical work on a) making a biodiesel and b) comparing it to fossil diesel. These can be either demonstrations or class practicals (pre-16 or post-16) depending on the ability level of the students and the time available;

▼ reference material which could be made into an information booklet on biodiesel, *Introducing biodiesel*, which can be for both pre-16 and post-16 students ;

▼ a comprehension exercise *Biodiesel and the environment*, with answers, for pre-16 students;

▼ a team exercise *Biodiesel – will you produce it?* to explore the social, environmental and economic aspects of biodiesel, suitable for pre-16 chemistry students; and

▼ worksheets covering the topics below for post-16 level, with answers.

    – alkenes;

    – infrared spectroscopy;

    – mass spectrometry;

    – calculations on biodiesel yields;

    – esters/biodiesel production; and

    – thermochemistry.

These can be tackled after some introductory work on biodiesel, such as reading the booklet *Introducing biodiesel* and/or the practical work.

This pack is the result of a Royal Society of Chemistry study visit to Chemoxy International, Middlesbrough in December 1995 and was based on original ideas produced by:

▼ Joe Burke, Tytherington County High School;

▼ Leslie Davidson, Keith Grammar School;

▼ John Milligan, Farr Secondary School;

▼ Jill Oldfield , Leeds Girls' High School; and

▼ Mary Richards, Arnold and Carlton College.

RS•C

Part 1 Pre-16 – Making biodiesel

Teacher's notes

The preparation of biodiesel from rape seed oil – or other suitable vegetable oil

Curriculum links

Biodiesel, a mixture of methyl esters of fatty acids, can be made very easily from a cooking oil made from rape seed, though other cooking oils may be tried. Enough fuel can be produced in a double lesson to burn, though it would not be pure enough to burn in an engine. This experiment could start off any fuel, thermochemistry or environmental topic as a demonstration, in which case it is probably better to allow the mixture from Stage 1 to react overnight as "one you have prepared earlier".

Level

Pre-16 chemistry/science students.

Timing

60 min.

Description

A cooking oil is mixed with methanol and a catalyst (potassium hydroxide). The resulting reaction (transesterification) produces biodiesel and glycerol (propane-1,2,3-triol) which separate into two layers. The biodiesel, in the top layer, is removed and then washed with water to remove potassium hydroxide.

Apparatus (per group)

▼　Access to a balance

▼　Access to a centrifuge

▼　One 250 cm³ conical flask

▼　Two 100 cm³ beakers

▼　One 10 cm³ measuring cylinder

▼　One 20 cm³ measuring cylinder

▼　Teat pipettes

▼　Centrifuge tubes

▼　Sample tube and label.

RS•C

Chemicals (per group)

▼ Deionised water;

▼ 100 g Rape seed oil or other vegetable oil – *eg* cooking oil

▼ 15 g Methanol

▼ 1 g 50% Potassium hydroxide solution.

Safety

▼ Wear eye protection

▼ Methanol is flammable and poisonous

▼ Potassium hydroxide is corrosive.

It is the responsibility of the teacher to carry out a risk assessment.

RS•C

RS•C

Biodiesel as a fuel

Curriculum links

This section can be used as part of the study of *Products from oil, the use of fuels and products of burning hydrocarbons.*

Comparing the fuels and using the data in the information booklet could provide a starting point for further *Experimental and investigative science* at pre-16 level. Viscosity, flash point and heat of combustion are possible areas of further investigation.

Level

Pre-16 chemistry/science students.

Timing

30 min.

Description

A sample of diesel and then a sample of biodiesel are burnt and the products of burning compared for sootiness and acidity.

Apparatus

▼ Glass tube with delivery tube, see below

▼ Filter pump

▼ Mineral wool and tweezers

▼ Glass filter funnel and rubber bung to fit

▼ Crucible

▼ Boiling tube with two tubes as shown below

▼ Pipette.

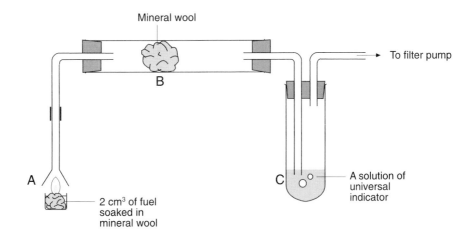

RS•C

Chemicals

▼   A few cm³ biodiesel (product from previous practical)

▼   A few cm³ paraffin (as used in commercial heaters) which is similar to diesel
       fuel

▼   A few cm³ universal indicator solution.

Safety

▼   Wear eye protection

▼   Paraffin is flammable

▼   It is the responsibility of the teacher to carry out a risk assessment.

RS·C

Introducing biodiesel
Teacher's notes

The infomation sheet, *Introducing biodiesel,* is written in straightforward language and is suitable for a range of students. It can be used as background material for both pre-16 and post-16 chemistry students as part of a fuel investigation. It can also be used with the question sheet *Biodiesel and the environment* as homework, or would suit self-study work as a relevant application of science. The last section of the booklet includes the chemistry of esterification and, though this is referred to in many of the post-16 question sheets it may be omitted for pre-16 students.

Answers to questions on biodiesel and the environment

1. A biofuel is a renewable fuel produced from plants.

2. Biodiesel is a biofuel produced from oil-producing plants.

3. A feedstock is the starting material for making a product. For biodiesel this could be rape, sunflowers or corn.

4. Set-aside land is land for which farmers are paid not to grow particular crops. (The government pays a subsidy to a farmer for land on which crops are no longer grown for use as food. This is aimed at, for example, reducing the European Union (EU) grain mountain).

5. Biodiesel is used in many parts of Europe and in the US.

6. Any suitable answer – *eg* environmental (waterways, inner cities).

7. a) It would take more land than is available to produce enough diesel to meet our present needs.

    b) One hectare produces 1 200 litres of diesel, so the lorry could travel 9 600 km.

8. The excise tax or VAT might be reduced.

9. a) After 7 days, 70% ±5% remains.

    b) After 14 days, 40% ± 5% remains.

RS•C

Team exercise: Biodiesel – will you produce it?

Teacher's notes

Curriculum links

This may be used as a data gatherng, data analysis and presentation exercise for 11–14 year old students.

Level

Pre-16 and particularly 11–14 year old students.

Timing

60 min if the exercise starts with a demonstration, plus 60 min for presentation work.

Description

In this exercise groups of students collect fact cards about biodiesel to decide whether it is worth producing. They then make a presentation to the rest of the class. It could follow on from a quick demonstration of the production of biodiesel from rape seed oil, where each stage has been prepared in advance and could be linked to a discussion of possible solutions to the future fossil fuel shortage.

Teaching tips

Each group could collect 10–15 cards so that each presentation is based on a different selection of facts. This should encourage debate after the presentation.

The teacher will require:

▼   a starting sheet, a summary sheet and 20 fact cards (which can be photocopied from the enclosed masters) for each group of students; and

▼   a *Biodiesel – will you produce it? –* teacher's fact indicator.

The exercise is in two parts.

1.   Collecting the facts

Every group is given the starting sheet *Biodiesel – will you produce it?* plus the summary sheet.

Each group is also given a different fact card to start them off.

The group looks at the card and writes down any facts it wishes to note in the matching place on the summary sheet. It then decides which further piece of information it wants by choosing one of the key words which are underlined on the the fact card.

The group then asks for a new card from the teacher by reference to one of the key words on the fact card. The teacher refers to the fact indicator to supply the correct fact card. (It works most smoothly if there is a set of 20 fact cards for each group.)

This is continued until each group has obtained 10–15 fact cards and the worksheets are filled in. This takes about 30 minutes.

2.   Presenting the facts

Each group of students now discusses the facts, decides whether or not to produce biodiesel and then gives a two minute presentation to the rest of the group. The presentation could include, posters, poems, songs, raps, *etc*. This can take at least another hour, including the final summing up by the teacher.

RS•C

Biodiesel – Teacher's fact indicator

Key words	Fact number
Biodiesel	1
Government influences	2
Rape	3
Chemical change	4
Uses	5
Chemical companies	6
Cost	7
Other European countries	8
Future	9
Fuel	10
Problems with biodiesel	11
Environment	12
Ordinary diesel	13
Scale	14
Cash crops	15
Renewable	16
More land	17
Alternatives	18
Diesel and water	19
Biodegradability	20

RS•C

Part 2 Post-16 – Making biodiesel

Teacher's notes

The preparation of biodiesel from rapeseed oil – or other suitable vegetable oil.

Curriculum links

Biodiesel, a mixture of methyl esters of fatty acids, can be made very easily from a cooking oil made from rape seed, though other cooking oils may be tried. Enough fuel can be produced in a double lesson to burn, though it would not be pure enough to burn in an engine. This experiment could be used as a starting point for further student investigations at post-16 level.

Timing

60 min.

Level

Post-16 chemistry students.

Description

A cooking oil is mixed with methanol and a catalyst (potassium hydroxide). The resulting reaction (transesterification) produces biodiesel and glycerol which separate into two layers. The biodiesel, in the top layer, is removed and then washed with water to remove potassium hydroxide.

Apparatus (per group)

▼ Access to a balance

▼ Access to a centrifuge (and magnetic stirrer if available)

▼ One 250 cm^3 conical flask

▼ Two 100 cm^3 beakers

▼ One 10 cm^3 measuring cylinder

▼ 20 cm^3 measuring cylinder

▼ Teat pipettes

▼ Centrifuge tubes

▼ Sample tube and label.

Chemicals (per group)

▼ Deionised water

▼ 200 g Rape seed oil or other vegetable oil – *eg* cooking oil

▼ 30 g Methanol

▼ 2 g 50% Potassium hydroxide solution.

▼ Wear eye protection

RS•C

▼   Methanol is flammable and poisonous

▼   Potassium hydroxide is corrosive.

It is the responsibility of the teacher to carry out a risk assessment.

Answers to questions on making biodiesel

1.   50% KOH has a concentration of 8.9 mol dm^{-3}.

2.   Glycerol (propane-1,2,3-triol) is in the lower layer.

3.   The washings remove potassium hydroxide.

4.   Appropriate calculation – *ie* commercially 1,200 kg rape seed oil gives
          1,100 kg of crude biodiesel. Therefore in this experiment you might expect to
          produce 200 x 1,100 g of crude biodiesel (= 183 g).

$$\frac{}{1,200}$$

Comparison of students' yield with 183 g.

RS•C

Post-16 worksheets

Teacher's notes

The question sheet which follows deals with some chemical principles which are related to the structure, manufacture and uses of biodiesel. The questions may be particularly useful for revision, in that they revise a number of topics via a different route which has an environmental theme. The booklet *Introducing biodiesel* is referred to in the worksheets marked with an asterisk and provides background reading for all of them.

Curriculum links

Alkenes* (students will need a databook giving infrared correlations.)

Infrared spectroscopy provides work on infrared spectra

Mass spectrometry

Calculations* provides work on (biodiesel) yields

The production of biodiesel provides work on esters

Thermochemistry*.

Answers to questions on post-16 worksheets

Alkenes

1.   The functional group for alkenes is C=C.

2.   Nine moles of H_2.

3.   Polyunsaturates contain many C=C.

4.   There is restricted rotation about C=C.

        Groups on the same side of the C=C are *cis*.

        Groups on different sides of the C=C are *trans*.

5.   Bromine is immediately decolourised.

        Electrophilic addition is occuring.

6.

7.   Main points:

▼   same volumes of each vegetable oil;

▼   titrate against bromine in hexane;

▼   end-point is when the drop of bromine is not decolourised; and

▼   compare volumes of bromine needed – the most unsaturated oil requires the most bromine.

Infrared spectroscopy

1.   The carbonyl of the ester group.

2.   Fossil diesel is a hydrocarbon, so no absorbance is seen for the ester group.

3.   This is the C–H stretch. All the molecules have a long hydrocarbon chain so this peak is present in all the spectra.

Mass Spectrometry

1.   (a)   An odd electron species/contains an unpaired electron.

    (b)   Positive ion.

2.   (a)   The positive ions are accelerated by negatively charged plates.

    (b)   The positive ions are deflected in a circular path by a magnetic field.

3.   The molecular ion peak is at mass (or, strictly, mass/charge) 32.

    Fragment ions include CH_3^+ and OH^+.

4.   These are caused by the presence of ^{13}C or ^2H.

RS•C

Calculations

1.

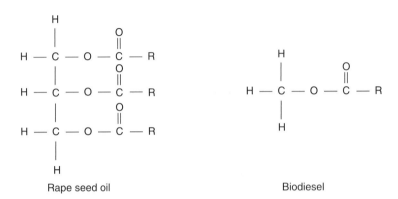

Rape seed oil   Biodiesel

where R is

Structures of rape seed oil and biodiesel

2. M_r values :

Rapeseed oil $(C_{57}H_{89}O_6)$ = 869

Methanol (CH_3OH) = 32

Biodiesel $(C_{19}H_{31}O_2)$ = 291

Glycerol $(C_3H_8O_3)$ = 92

3.   No. of moles of rapeseed oil = $\dfrac{1\,200\,000}{869}$ = 1380.9

4.   No. of moles of biodiesel = 3 x no. of moles of rapeseed oil

= 4142.7

5.   % yield   = $\dfrac{\text{mass of biodiesel} \times 100}{\text{theoretical mass}}$

= $\dfrac{1\,000\,000 \times 100}{4142.7 \times 291}$ = 83%

RS•C

RS•C

The production of biodiesel

1.

2. a) CH_3COOCH_3

   b) $HCOOCH_2CH_2CH_3$

   c) $HCOOCH_2CH_3$

3. Electronegativity is the ability of an atom to attract electrons to itself within a covalent bond.

4. An electron pair donor which forms bonds with electron-deficient carbon atoms.

5. Halogenoalkanes, aldehydes, ketones and acid chlorides *etc.*

6. Propane-1,2,3-triol.

7. Reagents: ethanoic acid or ethanoyl chloride or ethanoic anhydride and methanol.

   Conditions: concentrated sulphuric acid and reflux for ethanoic acid. Direct reaction for the others.

   Appropriate equation.

8.

Equation for the saponification of rape seed oil

RS•C

Thermochemistry

1. a)   The reaction is endothermic

    b)   C—O and O—H

    c)   C—O and O—H

    d) $\Delta H \approx 0$

2. $\Delta S \approx 0$

3. a) $\Delta G = \Delta H - T\Delta S$

    b) $\Delta G \approx 0$

4. a) $K \approx 1$

    b)   The reaction does not go to completion and an equilibrium mixture is formed.

The preparation of biodiesel from rape seed oil – or other suitable vegetable oils worksheet

Method

Stage 1

1. Weigh *ca* 100 g of rape seed oil into a conical flask.

2. Carefully:

    a)   add 15 g of methanol;

    b)   then slowly add 1 g of a 50% (50 g per 100 cm³ of solution) potassium hydroxide solution. Take care, potassium hydroxide is very corrosive.

    Adding the chemicals can be done directly into the conical flask on a top pan balance, zeroing the balance after each addition.

    Stir or swirl for 10 min.

Stage 2

1. Centrifuge the mixture for one minute (you will need several centrifuge tubes to deal with the quantity).

2. Decant off the top layers into a boiling tube and discard the lower layers.

3. Wash the product by adding 10 cm³ of distilled water to this top layer, with gentle mixing. Do not shake the mixture.

4. Repeat steps 1 and 2 once more.

5. Keep your product for further investigation.

Biodiesel as a fuel worksheet

Set up the apparatus shown below using first biodiesel and then fossil diesel as the fuel.

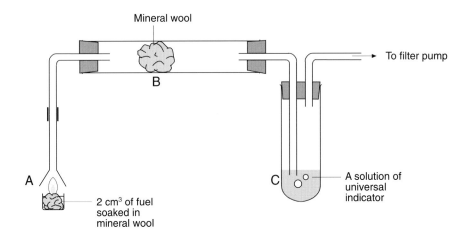

Burning fuels

Fill in your observations in the table. Record what happens at points A, B and C in the diagram.

	Biodiesel	Fossil diesel
A		
B		
C		

Write down your conclusions about differences in flammability, sootiness and acidity.

Extension

Suggest how your observations at B and C could be made quantitative.

Introducing biodiesel information sheets

Alternative Fuels

As the number of petrol- and diesel- powered vehicles on our roads increases, so too does the pollution caused by their exhaust emissions, especially in urban and built up areas.

Increased exhaust emissions are thought to be a factor in environmental problems. These include global warming through increasing carbon dioxide levels in the atmosphere and the formation of acid rain through emissions of sulfur dioxide and nitrogen oxides. The general health of those people living in built up areas may also be suffering.

The fuels we obtain from crude oil – such as petrol and diesel – will eventually run out and the race is on throughout the world to find alternative fuel sources. In Europe most of the interest lies in finding a replacement for *fossil diesel*, (diesel fuel derived from crude oil). It is important that any alternative fuel does not add to the environmental problems. Other alternatives like battery powered vehicles, which do not pollute their immediate environment, are only suitable for short journeys at relatively low speeds.

There are four recognised alternatives for the fossil fuels that we use in our vehicles.

Petrol (gasoline) alternatives	Diesel alternative
Methanol Ethanol Compressed natural gas (CNG)	Esterification of vegetable based oils – *eg* rape methyl ester (RME)

The alternatives to petrol – methanol, ethanol and CNG – could also be used to replace diesel, but extensive engine modifications would be needed. The only practical alternative for fossil diesel is *biodiesel* .

An internal combustion engine works by burning fuel in a small chamber. The expansion of the gases produced drives a piston which turns a shaft, which turns the wheels. Petrol engines need a spark to ignite the fuel, whereas diesel engines work by compressing the fuel, which heats it enough to cause ignition. So, the engine construction is different in each case.

A biofuel is a fuel made from a renewable source of growing vegetation. Ethanol, for example, can be produced by fermenting sugar, in which case it is a biofuel. When it is made from crude oil, as most of it is at present, it is not a biofuel.

What is biodiesel?

Biodiesel was first made in the 1940s. It is produced from a renewable source and is designed to replace the diesel used in diesel-powered vehicles.

The term biodiesel describes any biofuel produced from an oil-bearing vegetable feedstock. Biodiesel has been successfully produced from soya oil, sunflower oil, corn oil and rape seed oil but other vegetable oil-bearing crops might also be used. The specification (summary of the relevant properties for use as a fuel) of the biodiesel, varies depending on the feedstock used.

British Biodiesel Ltd uses rape seed oil to produce biodiesel. This is because rape grows well in the UK climate. You may have seen the very bright yellow crop growing in fields in spring and early summer.

The rape is harvested in July, August or September depending on the time of sowing. The collected seed is crushed to extract the rape seed oil, which is then reacted chemically to produce rape methyl ester (RME).

Main stages in biodiesel production

Stage 1   Growth and collection of rape seed

Rape seed is widely grown in the UK with high concentrations around the North East of England. Following recent changes in the European Union agricultural policy, farmers are under pressure to reduce food production through set-aside land. Rape seed is an economical crop that they may grow.

It takes 1 hectare (10 000 m^2, which is slightly less than the size of a football pitch) of land to grow enough rape to produce three tonnes (3000 kg) of rape seed.

Set-aside land is land for which farmers are paid by the government to stop them from growing on it particular crops for food. This means that they may grow food crops as long as the end use is not food!

Stage 2   Extraction of rape seed oil

The collected rape seed is crushed to extract rape seed oil. A by-product of this process is rape meal, a high protein animal feed. Rape meal could eventually replace imported soya bean-based animal feed.

From every hectare of rape harvested, about 1.2 tonnes of rape seed oil is extracted and 1.8 tonnes of animal feed produced. Approximately 1200 litres of biodiesel can be produced from the rape seed oil.

Stage 3   Production of rape methyl ester (biodiesel)

The final stage in the production of biodiesel involves a chemical change. This is called transesterification. Using potassium hydroxide as a catalyst, the rape seed oil is reacted with methanol (a type of alcohol) to produce biodiesel and glycerol (propane-1,2,3-triol). Glycerol is a valuable by-product. It is used in the motor industry as antifreeze and is refined for use in the pharmaceutical industry.

A catalyst speeds up a reaction without being used up.

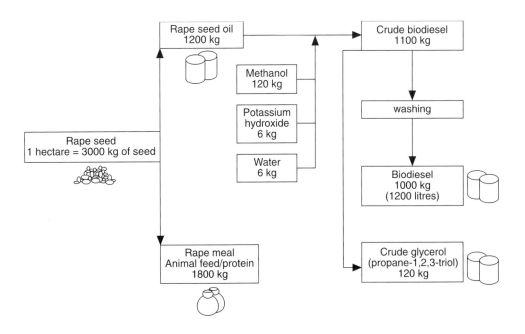

**Flow chart showing the stages of biodiesel production
(with approximate quantities)**

The history of biofuels

Although biodiesel may be a relatively new form of energy, biofuels have existed since the early 1900s. In 1914 Brazil successfully used ethanol manufactured from sugar cane, and in the past 20 years over 50% of the cars in Brazil have been running on an ethanol/petroleum mix.

The first diesel substitutes made from vegetable oils were produced in the US in the early 1950s.

Tests on exhaust emissions of vehicles using blends of biodiesel and fossil diesel in the ratio of 20:80 (biodiesel:fossil diesel) showed that particulate matter (smoke or soot), carbon monoxide levels and the total hydrocarbon levels were all significantly lower compared to those from fossil diesel. Biodiesel (made from soya oil) is used in most of the US.

The use of biofuels in Europe grew as the price of fossil diesel rose in the 1970s.

In France, biodiesel (based on sunflower oil) is available at the pump as a blend of 5% biodiesel with 95% fossil diesel At the moment there is no tax on biofuels.

In Italy, biodiesel is also produced from sunflower oil, as the crop is widely grown there.

Austria has been heavily involved in the production of biodiesel from rape seed oil for over 10 years and has produced the most extensive studies on rape methyl ester (RME). Although the chemistry of biodiesel based on sunflower oil is essentially the same as biodiesel based on rape seed oil, its specification is slightly different.

In the UK, British Biodiesel Ltd uses rape seed oil to produce biodiesel.

British Biodiesel Ltd

In January 1993 the East Durham Biodiesel working group was set up by a group of farmers to start the British Biodiesel industry.

In early 1994 the group formed a successful association with three other North East based companies to bring together representatives from the farming and the chemical industries. The three companies were approached for their specialist skills in the areas of seed collection, crushing and transesterification – the name for the chemical process that changes the rape seed oil to biodiesel.

The three company association consists of:

Farmway, Darlington a farmers cooperative involved with the farming and harvesting of rape;

Unitrition, Selby a specialised seed crushing company; and

Chemoxy International, Middlesbrough, a chemical manufacturer with the facilities for transesterification.

An increasing number of UK companies now use biodiesel.

Ecostatistics of UK biodiesel compared with fossil diesel

Soot and particulate emissions	Reduced by an average of 40%
Hydrocarbon and aromatic* emissions	Reduced by 63%
Sulfur content	Significantly less
Carbon dioxide	Little change
Carbon monoxide	Significantly less
NO$_x$ emissions	Slightly lower
Toxicity	Over 100 times less toxic. Oral (by mouth) and dermal (through the skin) LD$_{50}$s both exceed 2 g/kg
Biodegradability	Much better, more than 95% biodegrading in 21 days compared with around 50% for diesel. So in an accidental spill, the biodiesel would be digested by bacteria much faster than fossil diesel.

*Aromatic compounds are related to benzene, and can be harmful

LD$_{50}$ stands for lethal dose 50 and is the dose which would kill 50% of a test population of organisms. The larger the number the safer the substance. For example, if biodiesel were spilled in a river, it would take a dose of more than 2 g for each 1 kg of fish, for half the fish population to die. This would compare with half the population of fish being killed off with a dose of fossil diesel of as little as 1/100 to 1/200 g/kg.

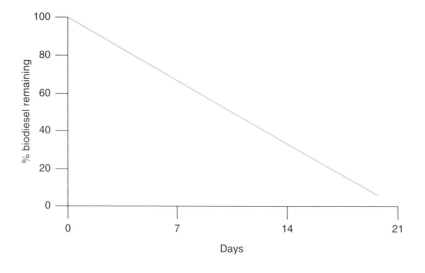

Biodegradability

Specifications

The following table compares some relevant properties of the specification for fossil diesel with those of the specification for a biodiesel.

Property	Fossil diesel	Biodiesel
Cetane index*	46–49	49
Density (g/cm^3)	0.820–0.860	0.860–0.890
Viscosity† (mm^2/s)	2.00–4.50	3.50–5.00
Flash point§ (°C)	55	100
Water content (mg/kg)	200	500 max
Sulphur content (% wt)	0.20	0.01 max
Particulates (g/m^3)	24	20 max
Calorific value (MJ/l)	35–36	33–34
Toxicity	high	low
Biodegradability	50% /21 days	95% /21 days

* The cetane rating is comparable to the octane rating that is used for petrol.
   Basically, the larger the number the more evenly the fuel burns in the engine.
†Viscosity measures how treacly the oil is. The larger the number, the slower is the flow.
§ The flash point is the temperature at which the fuel will self-ignite.

Replacement of fossil diesel with biodiesel in the UK

Biodiesel can be used as a replacement for fossil diesel without any engine modifications. Current users have noticed significant reductions in exhaust emissions and a general cleaner running performance with no detrimental effect or loss of acceleration.

British Biodiesel Ltd does not intend to compete with the fossil diesel made by the large petroleum refiners and so biodiesel will not be available to the general public for use in private cars. It will be sold where its ecological advantages are most useful. For example, the use of biodiesel on waterways for leisure craft and barges would reduce the effects caused by spillages of fossil diesel as it is so quickly broken down (95% in 21 days). Urban and inner city areas would significantly benefit from using biodiesel in buses and taxis. Other possible markets include the use of biodiesel in enclosed areas with the use of heavy plant equipment, fork lift trucks and off-shore work.

Even if every arable acre in the UK were to grow rape, it would still only produce between 7 and 10% of our requirements for diesel.

Costs

	Fossil Diesel pence /l	Biodiesel pence /l
Production prices	7.70	38.00
Distribution	6.57	3.94
Excise	30.70	30.74
	44.97	**72.64**
+ VAT @ 17.5%	7.87	12.71
	52.84	**85.35**
+ Retailer's mark up	4.14	4.14
Total	**56.98**	**89.49**

A comparison of fossil diesel and biodiesel retail price 1995

The chemistry of biodiesel production

All vegetable oils are large molecules with the following general form.

(R is a long chain hydrocarbon. The three R groups may be the same or different).
They have three molecules of long chain fatty acids and are known as triglycerides, because the stem is glycerol (propane -1,2,3-triol).

Glycerol

Fatty acids

Vegetable oils are *esters* of glycerol and fatty acids. They are called glyceride esters. There are many different fatty acids with different hydrocarbon chain lengths and degrees of unsaturation (number of carbon–carbon double bonds). The compositions of different oils (*eg* rape, olive, sunflower *etc*) and of individual types of oil from crops grown in different areas of the world are different and provide a means of identifying the source of the oil.

Biodiesel is produced by turning the glycerol esters into methyl esters. This is done by mixing the oil with an excess of methanol in the presence of potassium hydroxide (used as the catalyst). Methanol displaces glycerol which is then separated from the resulting methyl esters.

| Glycerol ester | Methanol | Glycerol | Methyl ester (biodiesel) |

For rape seed oil, one possibility for R is:

although there are others.

Biodiesel and the environment

Use the booklet *Introducing biodiesel* to answer the following questions.

1. What is a biofuel?

2. What is biodiesel?

3. What is a feedstock, and what are the feedstocks for biodiesel?

4. What is meant by set-aside land?

5. Where is biodiesel used at present?

6. Pick two advantages of replacing fossil diesel with biodiesel. For each of your advantages, say where the use might be particularly important. Explain your answer.

7. a) Explain why biodiesel cannot completely replace fossil diesel.

   b) How many miles could a lorry with a diesel consumption of 8 km/l travel on the biodiesel from a year's production from a field of rape of area one hectare?

8. Look at the 1995 prices of diesel and biodiesel. Suggest how biodiesel may be made more similar in price to diesel.

9. Use the graph *Biodegradability* to estimate what percentage of biodiesel would remain in the environment after

   a)   7 days, and

   b)   14 days?

Biodiesel – will you produce it? worksheet

Biodiesel is a fuel. Your company has to decide whether it would be good for the firm to produce large quantities of biodiesel fuel.

There are 20 facts to help your company decide the best thing to do. Your teacher will tell you how many facts you are allowed to collect.

You will be given a fact card and you then decide the next piece of information to collect by choosing one of the words in italics.

Fill in and use the summary sheet provided to gather all the facts together.

Once you have collected enough facts, your group has to produce a two minute presentation of its findings to the rest of the class. The following points should be answered in your presentation:

▼   whether you will produce the biodiesel or not;

▼   why you made this decision; and

▼   if anything might make you change your mind.

Summary sheet

Fact	Keypoints
1. Biodiesel	
2. Government influences	
3. Rape	
4. Chemical change	
5. Uses	
6. Chemical companies	
7. Cost	
8. Other European countries	
9. Future	
10. Fuel	
11. Problems with biodiesel	
12. Environment	
13. Ordinary diesel	
14. Scale	
15. Cash crops	
16. Renewable	
17. More land	
18. Alternatives	
19. Diesel and water	
20. Biodegradability	

Fact cards

Photocopy these facts onto separate pieces of card

✂ -

Biodiesel

FACT NUMBER 1

Biodiesel is a fuel that can be made from a plant called *rape*. The seeds of the rape plant are similar in many ways to sunflower seeds. Biodiesel could be the fuel of the *future*. It may solve many *environmental problems*.

✂ -

Government influences

FACT NUMBER 2

In 1995 the government taxed *biodiesel* at the same level as *ordinary diesel*. This added about 31p to the price per litre of biodiesel. In other European countries there is often less tax on biodiesel. The situation may change in the *future*.

✂ -

Rape

FACT NUMBER 3

Rape plants are grown by farmers as a *cash crop* often on set-aside land. This is land for which the government pays farmers not to grow certain food crops. The seeds are gathered and *chemically changed* into biodiesel. Around 1.5 million tonnes of rape is grown in the UK though more is grown in many other *European countries*.

✂ -

Chemical change

FACT NUMBER 4

The production of *biodiesel* involves farmers, *chemical companies* and customers. The farmers grow the rape seed and sell it to the chemical companies. The companies then produce the biodiesel by a chemical reaction which is fairly cheap and straightforward and gives a good yield. There are many *uses of biodiesel*.

Fact cards

Uses

FACT NUMBER 5

As a fuel, biodiesel is renewable and the *environmental facts* about biodiesel are interesting. The table gives more information about the energy content in biodiesel. It is interesting to compare biodiesel with *ordinary diesel*.

Fuel	Energy content	Efficiency
Biodiesel	34 MJ/l	40.7%

Biodiesel has many suitable markets where it would be particularly useful. These include national parks, theme parks, off-shore equipment, inner city buses and inland waterways.

Chemical companies

FACT NUMBER 6

As the chemical company you will have to *chemically change* the *rape* seed. Find out

▼ *how much it costs* to produce;
▼ the *uses* of biodiesel; and
▼ if there are any *problems associated with biodiesel*.

Cost

FACT NUMBER 7

The cost of one litre of biodiesel in 1995 was about 89p.
The cost is made up of:

▼ seeds;
▼ extraction of the oil;
▼ transport;
▼ *chemically changing* the *rape* seed;
▼ *government influences*; and
▼ other factors.

Other European countries

FACT NUMBER 8

In other European countries the *government influence* is often positive; the tax on *biodiesel* is low and the governments encourage *rape* seed production because biodiesel has *environmental advantages*.

Fact cards

FACT NUMBER 9

In the future many things might change:

- ▼   if the *government* changes the tax rate on biodiesel it should make the price of one litre more competitive;
- ▼   if the UK follows the example of other *European countries* which have lower taxes on biodiesel, it could have marked results for biodiesel production;
- ▼   if there is a war in any of the fossil fuel-producing countries the need for biodiesel as a fuel might increase; and
- ▼   if biodiesel were to be made on a larger *scale*.

FACT NUMBER 10

Fuels are store-houses of energy. When fuels burn in plenty of oxygen they normally make carbon dioxide and water along with lots of heat. Some fuels are *renewable*. If the fuel does not burn completely some dangerous gases may be produced which affect the *environment*.

FACT NUMBER 11

There are problems associated with biodiesel – *eg*:

- ▼   *rape* pollen may cause hay fever in some people;
- ▼   *diesel and water* do not mix;
- ▼   some people think that burning *biodiesel* gives off a smell rather like a chip shop;
- ▼   *chemically changing* the rape seed is expensive;
- ▼   growing rape requires a large scale use of land; and
- ▼   with biodiesel there is a slight increase in some *environmental* pollutants.

There are some *alternatives* to biodiesel.

FACT NUMBER 12

A green fuel is an environmentally friendly fuel, *ie* it causes little damage to the environment. Biodiesel compares very well with *ordinary diesel*. It is a renewable fuel and quickly *biodegrades*.
The following table compares the exhaust emissions from ordinary diesel and those from biodiesel.

	Ordinary diesel	Biodiesel
Nitrogen oxides	100	94.1
Carbon monoxide	100	65.4
Hydrocarbons	100	69.1
Carbon dioxide	100	100.3
Smoke	100	59

Fact cards

FACT NUMBER 13

Fossil diesel (ordinary diesel) was formed from partially decayed organisms that used to live in warm oceans millions of years ago. The organisms died, fell to the ocean bed in deep layers, were covered in mud, squashed and heated and eventually changed into crude oil. Ordinary diesel can be separated from crude oil by fractional distillation.

Ordinary diesel is a non-*renewable fuel* and there are some worries about it causing *environmental problems*. The table tells you some more about the energy in ordinary diesel.

Fuel	Energy content	Efficiency
Ordinary diesel	36 MJ/l	38.2%

FACT NUMBER 14

If the scale of *biodiesel* production could be increased, larger quantities could be made; it is cheaper to make a lot of a product than to make a small amount. This means that *more land* would be used for growing *rape* seed. A producer must be able to guarantee a supply of biodiesel to customers or they will not risk buying the biodiesel in the first place. It depends on what happens in the *future*.

FACT NUMBER 15

The plants that farmers grow for money are called cash crops. For farmers to grow more *rape* seed the *government influences* must be right. There must also be a *chemical company* to buy the rape seed so that the crop is profitable.

FACT NUMBER 16

Fossil *fuels* such as oil and coal take millions of years to form. This means that when we use the fuels they will not be replaced for millions of years. We call these fuels non-renewable. Biodiesel is a renewable fuel because it only takes a few months for the seeds to grow and the seeds can then be *chemically changed* into the fuel.

Fact cards

✂ -

More land

FACT NUMBER 17

The amount of biodiesel that can be made in the UK is limited and it is not intended to compete with the main markets of *ordinary diesel*. Biodiesel requires a lot of land (1 hectare of rape produces 1200 litres of biodiesel), so to have enough biodiesel to run all the diesel engines in the UK would be impossible. The main areas for biodiesel *use* are decided by the environmental factors and the by-products of *chemicallly changing* the rape seed.

✂ -

Alternatives

FACT NUMBER 18

Alternatives to biodiesel include:

▼ using sugar cane to produce alcohol;
▼ continued use of *ordinary diesel*;
▼ splitting water to make hydrogen gas;
▼ using battery-powered engines; and
▼ using compressed natural gas (a fossil *fuel*).

Each of the alternatives has advantages and disadvantages. Factors like *cost*, availability, technology, demand and pollution levels need to be considered.

✂ -

Diesel and water

FACT NUMBER 19

Diesel and water do not mix, so both spilt biodiesel and *ordinary diesel* form a thin layer on water through which no oxygen can pass. When diesel spills into water it can cause many problems – *eg*:

▼ fish die due to lack of oxygen;
▼ plants die due to lack of air; and
▼ birds can get covered in diesel and die due to poisoning and cold.

It takes time for the spilt diesel to break down naturally and we describe this as its *biodegradability*.

✂ -

Biodegradability

FACT NUMBER 20

The biodegradability of a *fuel* describes how long it takes to be broken down by bacteria in the environment. Spilled oil can cause problems so the faster it biodegrades the better. Twenty-one days after spilling ordinary diesel, 50% still remains, 21 days after spilling biodiesel, less than 5% still remains. These results may explain some of the proposed *uses* for biodiesel.

The preparation of biodiesel from rape seed oil – or other suitable vegetable oils

Method

Note

This method produces biodiesel relatively quickly, though the product is not pure enough to burn in an engine. It is more efficient to use a separating funnel after stage 1 and leave the mixture to separate. The washing separations are also better if each mixture is left for several hours to separate in a separating funnel (keeping the top layer each time). It is also better to have about five washings, but the whole process would then take a few days.

Stage 1

1. Weigh about 200 g of rape seed oil into a conical flask.

2. Carefully:

    a) add 30 g of methanol;

    b) then slowly add 2 g of a 50% (50 g per 100 cm^3 of solution) potassium hydroxide solution. Take care; potassium hydroxide is very corrosive.

    Additions of chemicals can be made directly into the conical flask on a top pan balance, zeroing the balance after each addition.

3. Mix well and leave overnight, stirring with a magnetic stirrer if possible.

Stage 2

1. Using portions of the mixture, centrifuge in tubes for 1 minute, then decant (or use a syringe) to remove the top layers into a conical flask. Discard the lower layers.

2. To wash the separated top layer, add 20 cm^3 of deionised water, with gentle mixing. Do not shake the mixture vigorously or an emulsion will form which is difficult to separate.

3. Repeat steps 1 and 2 once more.

4. The liquid you have is biodiesel. Weigh your product and keep it for further investigation.

Questions

1. What is the concentration in mol dm^{-3} of a 50% solution of potassium hydroxide?

2. What is left in the bottom layer in Stage 1?

3. What is the purpose of the washing steps?

4. In the commercial production of biodiesel, 1200 kg of rape seed oil produces 1100 kg of crude biodiesel. How does your yield compare with this?

Alkenes worksheet

Rape seed oil contains approximately 20% of polyunsaturated oils, 10% of saturated oils and 70% monounsaturated oils.

Taking the structure of rape seed oil to be that below, answer the following questions:

1. What is the functional group for alkenes?

2. How many moles of hydrogen would be required to saturate one mole of rape seed oil?

3. Explain why rape seed oil is classed as a polyunsaturated oil?

4. Geometrical isomerism is shown by rape seed oil. Using but-2-ene as an example, explain what is meant by geometrical isomerism.

5. What would be observed if bromine in hexane is added to rape seed oil? What type of reaction is occuring?

6. Give the mechanism for the reaction of bromine in hexane with ethene.

Extension

7. Devise an experiment to compare the unsaturation in samples of various vegetable oils.

Glycerol ester

Infrared spectroscopy worksheets

Infrared spectroscopy is a qualitative technique which helps to determine the types of bonds present in a variety of compounds. The infrared radiation is absorbed by molecules as specific covalent bonds or parts of the molecule bend or stretch. The wavenumber at which absorption takes place indicates the type of bond present in the sample.

Use a data book and the infrared spectra of rape seed oil, biodiesel and fossil diesel provided to answer the questions that follow the spectra.

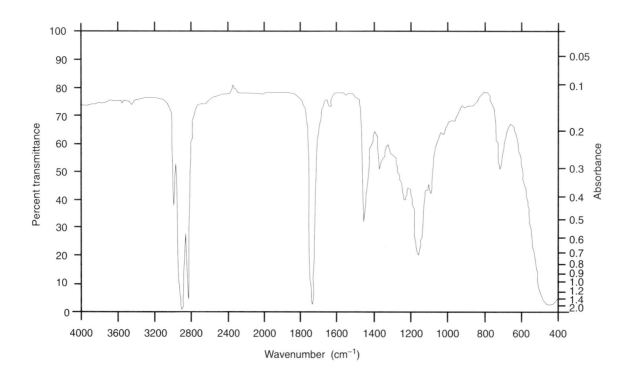

Infrared spectrum of rape seed oil

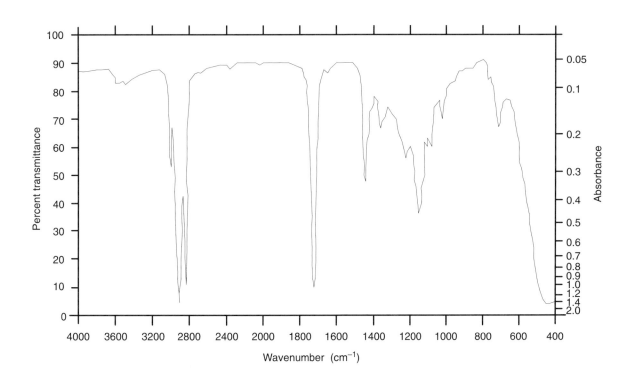

Infrared spectrum of biodiesel

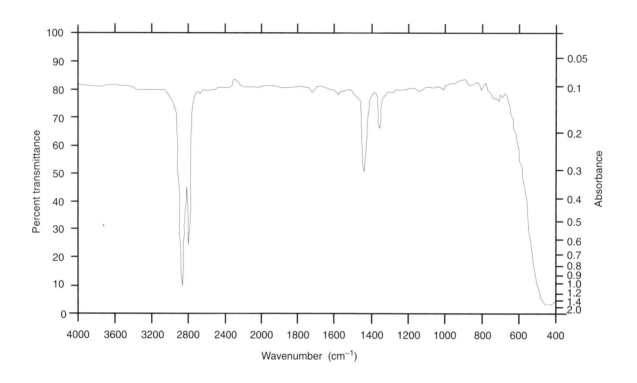

Infrared spectrum of diesel

Questions

1. In the infrared spectra for rape seed oil and biodiesel what does the peak at 1740 cm^{-1} indicate?

2. Why does fossil diesel not show a peak in this region of the spectrum?

3. Why do all the spectra show peaks at 2820–2860 cm^{-1}?

Mass spectrometry worksheet

Mass spectrometry is an analytical technique which can be used to help identify a variety of compounds. The molecules of vapourised material are bombarded by electrons, causing the formation of positive molecular ions, which may then fragment forming other cations. These ions are accelerated and then deflected before being detected in turn.

1. Molecular ions are radical cations. What is meant by (a) a radical and (b) a cation?

2. How are the cations (a) accelerated and (b) deflected?

3. In a mass spectrum of methanol where would you expect to find the molecular ion peak? Suggest two fragment ions that might be produced when this molecular ion fragments.

4. In many organic mass spectra a small peak is observed at a mass one unit greater than the molecular ion peak. Why is this?

Calculations worksheet

Use the equation provided in *Introducing biodiesel* and the flow chart below showing the stages in biodiesel production to answer the following questions.

1. Draw the structural formulae of rape seed oil and biodiesel.

2. Calculate the relative molecular masses of rape seed oil, methanol, biodiesel and glycerol (propane-1,2,3-triol).

3. 1200 kg of rapeseed oil is extracted from 3000 kg of rape seed. How many moles of oil are present?

4. Methanol is added in excess. How many moles of biodiesel should be produced?

5. In fact, 1200 kg of rape seed oil produces 1000 kg of biodiesel. What is the percentage yield of biodiesel?

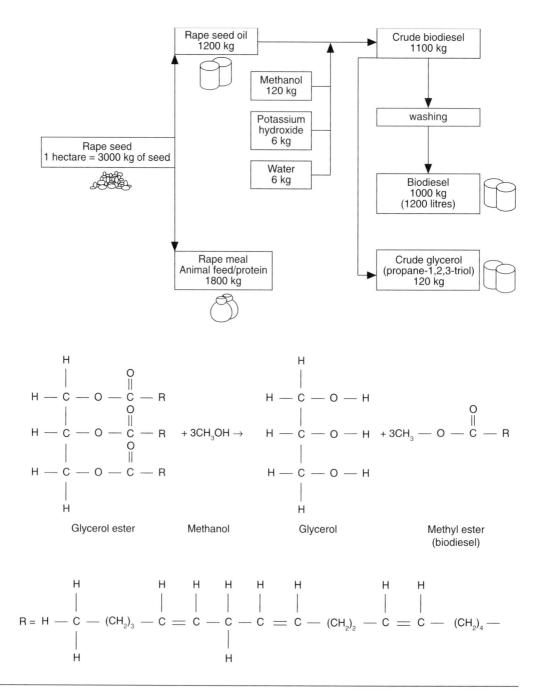

The production of biodiesel worksheet

Rape seed oil is the ester of a fatty acid. The ester functional group has a slight δ+ charge on the carbon of the C=O, as the oxygen is more electronegative than the carbon. Methanol acts as a nucleophile, attacking the electron deficient carbon, in a transesterification reaction (one ester being changed into a different ester).

1. Draw the functional group for esters.

2. Draw the structures of the following esters:

    (a) methyl ethanoate

    (b) propyl methanoate

    (c) ethyl methanoate.

    (If model kits are available you might like to make models of these esters)

3. What is the definition of electronegativity?

4. What is a nucleophile?

5. As well as esters what other types of organic chemicals are attacked by nucleophiles?

6. What is the systematic name for glycerol?

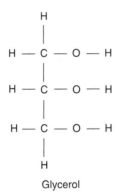
Glycerol

7. What reagent(s) and conditions would you use to produce methyl ethanoate? Write an equation for the reaction.

Extension

8. Many oils are obtained from plants and can be used to produce soaps, such as 'Palmolive'. This is known as saponification. Write an equation for the saponification of rape seed oil.

Thermochemistry worksheet

Using the equation below on the transesterification of rape seed oil answer the following questions.

1. a)   When bonds are broken, is the process exo- or endothermic?

    b)   Which bonds are broken for the reaction above to take place?

    c)   Which bonds are formed in the process?

    d)   What does this suggest about the value of the enthalpy change, ΔH, for the reaction?

2.   By considering the reactants and products, what can you say about the probable values of ΔS for the reaction?

3. a)   Write down the free energy equation in terms of ΔG, ΔH, ΔS and T.

    b)   What is the likely value of ΔG for the reaction?

4. a)   What does the value of ΔG suggest about the value of the equilibrium constant for the reaction?

    b)   What does this tell you about the reaction?

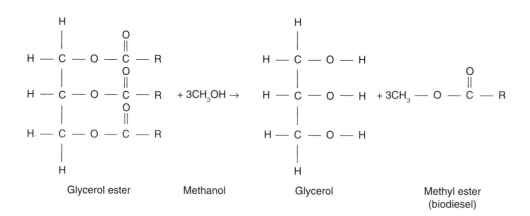

Glycerol ester   Methanol   Glycerol   Methyl ester (biodiesel)